Ever Since Darwin

Reflections in Natural History

BY STEPHEN JAY GOULD IN
NORTON PAPERBACK

EVER SINCE DARWIN
Reflections in Natural History

THE PANDA'S THUMB
More Reflections in Natural History

THE MISMEASURE OF MAN

HEN'S TEETH AND HORSE'S TOES
Further Reflections in Natural History

THE FLAMINGO'S SMILE
Reflections in Natural History

AN URCHIN IN THE STORM
Essays about Books and Ideas

WONDERFUL LIFE
The Burgess Shale and the Nature of History

BULLY FOR BRONTOSAURUS
Reflections in Natural History

FINDERS, KEEPERS
Treasures and Oddities of Natural History
Collectors from Peter the Great to Louis Agassiz
(with R. W. Purcell)

THE BOOK OF LIFE
An Illustrated History of Life on Earth
(general editor)

EIGHT LITTLE PIGGIES
Reflections in Natural History

Ever Since Darwin

Reflections in Natural History

Stephen Jay Gould

W · W · NORTON & COMPANY · NEW YORK · LONDON

Printed in the United States of America.

First published as a Norton paperback 1979; reissued 1992.

Library of Congress Cataloging in Publication Data
Gould, Stephen Jay. Ever Since Darwin.
Bibliography: p
Includes index.
1. Evolution—History. 2. Natural Selection—
History. I. Title.
QH361.G65 1977 575.01'62 77-22504

Designed by Nora Sheehan

ISBN 0-393-30818-9

W. W. Norton & Company, Inc.
500 Fifth Avenue, New York, N.Y. 10110
www.wwnorton.com

W. W. Norton & Company Ltd.
Castle House, 75/76 Wells Street, London W1T 3QT

4 5 6 7 8 9 0

For My Father

Who took me to see the
Tyrannosaurus when I was five

Contents

Ever Since Darwin

Reflections in Natural History

Prologue

"ONE HUNDRED YEARS without Darwin are enough," grumbled the noted American geneticist H. J. Muller in 1959. The remark struck many listeners as a singularly inauspicious way to greet the centenary of the *Origin of Species,* but no one could deny the truth expressed in its frustration.

Why has Darwin been so hard to grasp? Within a decade, he convinced the thinking world that evolution had occurred, but his own theory of natural selection never achieved much popularity during his lifetime. It did not prevail until the 1940s, and even today, though it forms the core of our evolutionary theory, it is widely misunderstood, misquoted, and misapplied. The difficulty cannot lie in complexity of logical structure, for the basis of natural selection is simplicity itself —two undeniable facts and an inescapable conclusion:

1. Organisms vary, and these variations are inherited (at least in part) by their offspring.

2. Organisms produce more offspring than can possibly survive.

3. On average, offspring that vary most strongly in directions favored by the environment will survive and propagate. Favorable variation will therefore accumulate in populations by natural selection.

These three statements do ensure that natural selection will operate, but they do not (by themselves) guarantee for it the fundamental role that Darwin assigned. The essence of

Darwin's theory lies in his contention that natural selection is the creative force of evolution—not just the executioner of the unfit. Natural selection must construct the fit as well; it must build adaptation in stages by preserving, generation after generation, the favorable part of a random spectrum of variation. If natural selection is creative, then our first statement on variation must be amplified by two additional constraints.

First, variation must be random, or at least not preferentially inclined toward adaptation. For, if variation comes prepackaged in the right direction, then selection plays no creative role, but merely eliminates the unlucky individuals who do not vary in the appropriate way. Lamarckism, with its insistence that animals respond creatively to their needs and pass acquired traits to offspring, is a non-Darwinian theory on this account. Our understanding of genetic mutation suggests that Darwin was right in maintaining that variation is not predirected in favorable ways. Evolution is a mixture of chance and necessity—chance at the level of variation, necessity in the working of selection.

Secondly, variation must be small relative to the extent of evolutionary change in the foundation of new species. For if new species arise all at once, then selection only has to remove former occupants to make way for an improvement that it did not manufacture. Again, our understanding of genetics encourages Darwin's view that small mutations are the stuff of evolutionary change.

Thus, Darwin's apparently simple theory is not without its subtle complexities and additional requirements. Nonetheless, I believe that the stumbling block to its acceptance does not lie in any scientific difficulty, but rather in the radical philosophical content of Darwin's message—in its challenge to a set of entrenched Western attitudes that we are not yet ready to abandon. First, Darwin argues that evolution has no purpose. Individuals struggle to increase the representation of their genes in future generations, and that is all. If the world displays any harmony and order, it arises only as an incidental result of individuals seeking their own advantage—the economy of Adam Smith transferred to nature. Second, Darwin maintained that evolution has no direction; it does not lead inevitably to higher things. Organisms become better adapted to their local environments, and that is all. The "degeneracy" of a parasite is as perfect as the gait of a gazelle. Third, Darwin applied a consistent philosophy of materialism to his interpretation of nature. Matter is the ground of all existence; mind, spirit, and God as well, are just words that express the wondrous results of neuronal complexity. Thomas Hardy, speaking for nature, expressed his distress at the claim that purpose, direction, and spirit had been banished:

> When I took forth at dawning, pool,
> Field, flock, and lonely tree,
> All seem to gaze at me
> Like chastened children sitting silent in a school;
>
> Upon them stirs in lippings mere
> (As if once clear in call,
> But now scarce breathed at all)—
> "We wonder, ever wonder, why we find us here!"

Yes, the world has been different ever since Darwin. But no less exciting, instructing, or uplifting; for if we cannot find purpose in nature, we will have to define it for ourselves. Darwin was not a moral dolt; he just didn't care to fob off upon nature all the deep prejudices of Western thought. Indeed, I suggest that the true Darwinian spirit might salvage our depleted world by denying a favorite theme of Western arrogance—that we are meant to have control and dominion over the earth and its life because we are the loftiest product of a preordained process.

In any case, we must come to terms with Darwin. And to do this, we must understand both his beliefs and their implications. All the disparate essays of this book are devoted to the exploration of "this view of life"—Darwin's own term for his new evolutionary world.

These essays, written from 1974–77, originally appeared in my monthly column for *Natural History Magazine,* entitled "This View of Life." They range broadly from planetary and geological to social and political history, but they are united (in my mind at least) by the common thread of evolutionary theory—Darwin's version. I am a tradesman, not a polymath; what I know of planets and politics lies at their intersection with biological evolution.

I am not unmindful of the journalist's quip that yesterday's paper wraps today's garbage. I am also not unmindful of the outrages visited upon our forests to publish redundant and incoherent collections of essays; for, like Dr. Seuss' Lorax, I like to think that I speak for the trees. Beyond vanity, my only excuses for a collection of these essays lie in the observation that many people like (and as many people despise) them, and that they seem to cohere about a common theme—Darwin's evolutionary perspective as an antidote to our cosmic arrogance.

The first section explores Darwin's theory itself, especially the radical philosophy that inspired H. J. Muller's complaint. Evolution is purposeless, nonprogressive, and materialistic. I approach the heavy message through some entertaining riddles: who was the Beagle's naturalist (not Darwin); why didn't Darwin use the word "evolution"; and why did he wait twenty-one years to publish his theory?

The application of Darwinism to human evolution forms the second section. I try to stress both our uniqueness and our unity with other creatures. Our uniqueness arises from the operation of ordinary evolutionary processes, not from any predisposition toward higher things.

In the third section, I explore some complex issues in evolutionary theory through their application to peculiar organisms. On one level, these essays are about deer with giant antlers, flies that eat their mother from inside, clams that evolve a decoy fish on their rear end, and bamboos that only flower every 120 years. On another level, they treat the issues of adaptation, perfection, and apparent senselessness.

The fourth section extends evolutionary theory to an exploration of patterns in the history of life. We find no story of stately progress, but a world punctuated with periods of mass extinction and rapid origination among long stretches of relative tranquility. I focus upon the two greatest punctuations—the Cambrian "explosion" that ushered in most complex animal life about 600 million years ago, and the Permian extinction that wiped out half the families of marine invertebrates 225 million years ago.

From the history of life, I move to the history of its abode, our earth (fifth section). I discuss both the ancient heros (Lyell) and the modern heretics (Velikovsky) who wrestled with the most general questions of all—does geological history have a direction; is change slow and stately, or rapid and cataclysmic; how does the history of life map the history of the earth? I find a potential resolution to some of these questions in the "new geology" of plate tectonics and continental drift.

The sixth section attempts to be comprehensive by looking in the small. I take a single, simple principle—the influence of size itself upon the shapes of objects—and argue that it applies to an astonishingly broad range of developmental phenomena. I include the evolution of planetary surfaces, the brains of vertebrates and the characteristic differences in shape between small and large medieval churches.

The seventh section may strike some readers as a break in the sequence. I have built laboriously from general principles down to their specific applications, and up again to their working in major patterns for life and the earth. Now I move to the history of evolutionary thought, particularly to the impact of social and political views upon supposedly "objective" science. But I see it as more of the same—another needle in scientific arrogance, with an added political message. Science is no inexorable march to truth, mediated by the collection of objective information and the destruction of ancient superstition. Scientists, as ordinary human beings, unconsciously reflect in their theories the social and political constraints of their times. As privileged members of society, more often than not they end up defending existing social arrangements as biologically foreordained. I discuss the general message in an obscure debate within eighteenth century embryology, Engels's views on human evolution, Lombroso's theory of innate criminality, and a twisted tale from the catacombs of scientific racism.

The final section pursues the same theme, but applies it to

contemporary discussions of "human nature"—the major impact of misused evolutionary theory upon current social policy. The first subsection criticizes as political prejudice the biological determinism that has recently deluged us with killer apes as ancestors, innate aggression and territoriality, female passivity as the dictate of nature, racial differences in IQ, etc. I argue that no evidence supports any of these claims, and that they represent just the latest incarnation of a long and sad story in Western history—blaming the victim with a stamp of biological inferiority, or using "biology as an accomplice," as Condorcet put it. The second subsection treats both my pleasure and unhappiness with the recently christened study of "Sociobiology," and its promise of a new, Darwinian account of human nature. I suggest that many of its specific claims are unsupported speculations in the determinist mode, but I find great value in its Darwinian explanation of altruism—as support for my alternate preference that inheritance has given us flexibility, not a rigid social structure ordained by natural selection.

These essays have suffered only minor alteration from their original status as columns in *Natural History Magazine*—errors corrected, parochialisms eliminated, and information updated. I have tried to attack the bugbear of essay collections, redundancy, but have retreated when my editorial knife threatened the coherence of any individual piece. At least I never use the same quote twice. Finally, my thanks and affection for editor-in-chief Alan Ternes, and for his copy editors Florence Edelstein and Gordon Beckhorn. They have supported me through a rash of cranky letters, and have shown the finest forebearance and discretion by using the lightest of editorial hands. Blame Alan, however, for all the really catchy titles—particularly for the sigmoid fraud of essay 15.

Sigmund Freud expressed as well as anyone the ineradicable impact of evolution upon human life and thought when he wrote:

> Humanity has in course of time had to endure from the hands of science two great outrages upon its naive self-

> love. The first was when it realized that our earth was not the center of the universe, but only a speck in a world-system of a magnitude hardly conceivable. . . . The second was when biological research robbed man of his particular privilege of having been specially created, and relegated him to a descent from the animal world.

I submit that the knowledge of this relegation is also our greatest hope for continuity on a fragile earth. May "this view of life" flower during its second century and help us to comprehend both the limits and the lessons of scientific understanding—as we, like Hardy's fields and trees, continue to wonder why we find us here.

1 | Darwiniana

1 | Darwin's Delay

FEW EVENTS INSPIRE more speculation than long and unexplained pauses in the activities of famous people. Rossini crowned a brilliant operatic career with *William Tell* and then wrote almost nothing for the next thirty-five years. Dorothy Sayers abandoned Lord Peter Wimsey at the height of his popularity and turned instead to God. Charles Darwin developed a radical theory of evolution in 1838 and published it twenty-one years later only because A. R. Wallace was about to scoop him.

Five years with nature aboard the *Beagle* destroyed Darwin's faith in the fixity of species. In July, 1837, shortly after the voyage, he started his first notebook on "transmutation." Already convinced that evolution had occurred, Darwin sought a theory to explain its mechanism. After much preliminary speculation and a few unsuccessful hypotheses, he achieved his central insight while reading an apparently unrelated work for recreation. Darwin later wrote in his autobiography:

> In October 1838 . . . I happened to read for amusement Malthus on *Population,* and being well prepared to appreciate the struggle for existence which everywhere goes on from long continued observation of the habits of animals and plants, it at once struck me that under these circumstances favorable variations would tend to be preserved and unfavorable ones to be destroyed. The result of this would be the formation of new species.

Darwin had long appreciated the importance of artificial selection practiced by animal breeders. But until Malthus's vision of struggle and crowding catalyzed his thoughts, he had not been able to identify an agent for natural selection. If all creatures produced far more offspring than could possibly survive, then natural selection would direct evolution under the simple assumption that survivors, on the average, are better adapted to prevailing conditions of life.

Darwin knew what he had achieved. We cannot attribute his delay to any lack of appreciation for the magnitude of his accomplishment. In 1842 and again in 1844 he wrote out preliminary sketches of his theory and its implications. He also left strict instructions with his wife to publish these alone of his manuscripts if he should die before writing his major work.

Why then did he wait for more than twenty years to publish his theory? True, the pace of our lives today has accelerated so rapidly—leaving among its victims the art of conversation and the game of baseball—that we may mistake a normal period of the past for a large slice of eternity. But the span of a man's life is a constant measuring stick; twenty years is still half a normal career—a large chunk of life even by the most deliberate Victorian standards.

Conventional scientific biography is a remarkably misleading source of information about great thinkers. It tends to depict them as simple, rational machines pursuing their insights with steadfast devotion, under the drive of an internal mechanism subject to no influence but the constraints of objective data. Thus, Darwin waited twenty years—so the usual argument runs—simply because he had not completed his work. He was satisfied with his theory, but theory is cheap. He was determined not to publish until he had amassed an overwhelming dossier of data in its support, and this took time.

But Darwin's activities during the twenty years in question display the inadequacy of this traditional view. In particular, he devoted eight full years to writing four large volumes on the taxonomy and natural history of barnacles. Before this single fact, the traditionalists can only offer pap—something like: Darwin felt that he had to understand species thoroughly before proclaiming how they change; this he could do only by working out for himself the classification of a difficult group of organisms—but not for eight years, and not while he sat on the most revolutionary notion in the history of biology. Darwin's own assessment of the four volumes stands in his autobiography.

> Besides discovering several new and remarkable forms, I made out the homologies of the various parts . . . and I proved the existence in certain genera of minute males complemental to and parasitic on the hermaphrodites. . . . Nevertheless, I doubt whether the work was worth the consumption of so much time.

So complex an issue as the motivation for Darwin's delay has no simple resolution, but I feel sure of one thing: the negative effect of fear must have played at least as great a role as the positive need for additional documentation. Of what, then, was Darwin afraid?

When Darwin achieved his Malthusian insight, he was twenty-nine years old. He held no professional position, but he had acquired the admiration of his colleagues for his astute work aboard the *Beagle.* He was not about to compromise a promising career by promulgating a heresy that he could not prove.

What then was his heresy? A belief in evolution itself is the obvious answer. Yet this cannot be a major part of the solution; for, contrary to popular belief, evolution was a very common heresy during the first half of the nineteenth century. It was widely and openly discussed, opposed, to be sure, by a large majority, but admitted or at least considered by most of the great naturalists.

An extraordinary pair of Darwin's early notebooks may contain the answer (see H. E. Gruber and P. H. Barrett, *Darwin on Man,* for text and extensive commentary). These so-called M and N notebooks were written in 1838 and 1839, while Darwin was compiling the transmutation notebooks that formed the basis for his sketches of 1842 and 1844. They contain his thoughts on philosophy, esthetics, psychology, and anthropology. On rereading them in 1856, Darwin described them as "full of metaphysics on morals." They include many statements showing that he espoused but feared to expose something he perceived as far more heretical than evolution itself: philosophical materialism—the postulate that matter is the stuff of all existence and that all mental and spiritual phenomena are its by-products. No notion could be more upsetting to the deepest traditions of Western thought than the statement that mind—however complex and powerful—is simply a product of brain. Consider, for example, John Milton's vision of mind as separate from and superior to the body that it inhabits for a time (*Il Penseroso,* 1633).

> Or let my lamp, at midnight hour,
> Be seen in some high lonely tower,
> Where I may oft outwatch the Bear,
> With thrice-great Hermes,[1] or unsphere
> The spirit of Plato, to unfold
> What worlds or what vast regions hold
> The immortal mind that hath forsook
> Her mansion in this fleshly nook.

The notebooks prove that Darwin was interested in philosophy and aware of its implications. He knew that the primary feature distinguishing his theory from all other evolutionary doctrines was its uncompromising philosophical materialism. Other evolutionists spoke of vital forces, directed history, organic striving, and the essential irreducibil-

---

1 | "The Bear" refers to the constellation of Ursa major (the Great Bear), better known to us by its tail and hindquarters—the big dipper. "Thrice great Hermes" is Hermes Trismegistus (a Greek name for Thoth, Egyptian god of wisdom). The "hermetic books," supposedly authored by Thoth, are a collection of metaphysical and magical works that exerted great influence in seventeenth century England. They were equated by some with the Old Testament as a parallel source of pre-Christian wisdom. They waned in importance when exposed as a product of Alexandrian Greece, but survive in various doctrines of the Rosicrucians, and in our phrase "hermetic seal."

ity of mind—a panoply of concepts that traditional Christianity could accept in compromise, for they permitted a Christian God to work by evolution instead of creation. Darwin spoke only of random variation and natural selection.

In the notebooks Darwin resolutely applied his materialistic theory of evolution to all phenomena of life, including what he termed "the citadel itself"—the human mind. And if mind has no real existence beyond the brain, can God be anything more than an illusion invented by an illusion? In one of his transmutation notebooks, he wrote:

> Love of the deity effect of organization, oh you materialist! . . . Why is thought being a secretion of brain, more wonderful than gravity a property of matter? It is our arrogance, our admiration of ourselves.

This belief was so heretical that Darwin even sidestepped it in *The Origin of Species* (1859), where he ventured only the cryptic comment that "light will be thrown on the origin of man and his history." He gave vent to his beliefs, only when he could hide them no longer, in the *Descent of Man* (1871) and *The Expression of the Emotions in Man and Animals* (1872). A. R. Wallace, the codiscoverer of natural selection, could never bring himself to apply it to the human mind, which he viewed as the only divine contribution to the history of life. Yet Darwin cut through 2,000 years of philosophy and religion in the most remarkable epigram of the M notebook:

> Plato says in *Phaedo* that our "imaginary ideas" arise from the preexistence of the soul, are not derivable from experience—read monkeys for preexistence.

In his commentary on the M and N notebooks, Gruber labels materialism as "at that time more outrageous than evolution." He documents the persecution of materialistic beliefs during the late eighteenth and early nineteenth century and concludes:

> In virtually every branch of knowledge, repressive methods were used: lectures were proscribed, publication was hampered, professorships were denied, fierce invective and ridicule appeared in the press. Scholars and scien-

tists learned the lesson and responded to the pressures on them. The ones with unpopular ideas sometimes recanted, published anonymously, presented their ideas in weakened forms, or delayed publication for many years.

Darwin had experienced a direct example as an undergraduate at the University of Edinburgh in 1827. His friend W. A. Browne read a paper with a materialistic perspective on life and mind before the Plinian Society. After much debate, all references to Browne's paper, including the record (from the previous meeting) of his intention to deliver it, were expunged from the minutes. Darwin learned his lesson, for he wrote in the M notebook:

> To avoid stating how far, I believe, in Materialism, say only that emotions, instincts, degrees of talent, which are hereditary are so because brain of child resembles parent stock.

The most ardent materialists of the nineteenth century, Marx and Engels, were quick to recognize what Darwin had accomplished and to exploit its radical content. In 1869, Marx wrote to Engels about Darwin's *Origin:*

> Although it is developed in the crude English style, this is the book which contains the basis in natural history for our view.

A common bit of folklore—that Marx offered to dedicate volume 2 of *Das Kapital* to Darwin (and that Darwin refused)—turns out to be false. But Marx and Darwin did correspond, and Marx held Darwin in very high regard. (I have seen Darwin's copy of *Das Kapital* in his library at Down House. It is inscribed by Marx who calls himself a "sincere admirer" of Darwin. Its pages are uncut. Darwin was no devotee of the German language.)

Darwin was, indeed, a gentle revolutionary. Not only did he delay his work for so long, but he also assiduously avoided any public statement about the philosophical implications of his theory. In 1880, he wrote:

> It seems to me (rightly or wrongly) that direct arguments against Christianity and Theism hardly have any effect

> on the public; and that freedom of thought will best be promoted by that gradual enlightening of human understanding which follows the progress of science. I have therefore always avoided writing about religion and have confined myself to science.

Yet the content of his work was so disruptive to traditional Western thought that we have yet to encompass it all. Arthur Koestler's campaign against Darwin, for example, rests upon a reluctance to accept Darwin's materialism and an ardent desire once again to invest living matter with some special property (see *The Ghost in the Machine* or *The Case of the Midwife Toad*). This, I confess, I do not understand. Wonder and knowledge are both to be cherished. Shall we appreciate any less the beauty of nature because its harmony is unplanned? And shall the potential of mind cease to inspire our awe and fear because several billion neurons reside in our skulls?

2 Darwin's Sea Change, or Five Years at the Captain's Table

GROUCHO MARX ALWAYS delighted audiences with such outrageously obvious questions as "Who's buried in Grant's tomb?" But the apparently obvious can often be deceptive. If I remember correctly, the answer to who framed the Monroe Doctrine? is John Quincy Adams. Most biologists would answer "Charles Darwin" when asked, "Who was the naturalist aboard the H.M.S. *Beagle?*" And they would all be wrong. Let me not sound too shocking at the outset. Darwin was on the *Beagle* and he did devote his attention to natural history. But he was brought on board for another purpose, and the ship's surgeon, Robert McKormick, originally held the official position of naturalist. Herein lies a tale; not just a nit-picking footnote to academic history, but a discovery of some significance. Anthropologist J. W. Gruber reported the evidence in "Who Was the *Beagle*'s Naturalist?" written in 1969 for the *British Journal for the History of Science.* In 1975, historian of science H. L. Burstyn attempted to answer the obvious corollary: If Darwin wasn't the *Beagle*'s naturalist, why was he on board?

No document specifically identifies McKormick as an official naturalist, but the circumstantial evidence is overwhelming. The British navy, at the time, had a well-established tradition of surgeon-naturalists, and McKormick had deliberately educated himself for such a role. He was an adequate, if not brilliant, naturalist and performed his tasks with distinction on other voyages, including Ross's Antarctic expedition (1839–1843) to locate the position of the South Magnetic Pole. Moreover, Gruber has found a letter from the Edinburgh naturalist Robert Jameson addressed to "My dear Sir" and full of advice to the *Beagle*'s naturalist on collection and preservation of specimens. In the traditional view, no one but Darwin himself could have been the recipient. Fortunately, the name of the addressee is on the original folio. It was written to McKormick.

Darwin, to cut the suspense, sailed on the *Beagle* as a companion to Captain Fitzroy. But why would a British captain want to take as a companion for a five-year journey a man he had only met the previous month? Two features of naval voyages during the 1830s must have set Fitzroy's decision. First of all, voyages lasted for many years, with long stretches between ports and very limited contact by mail with friends and family at home. Secondly (and however strange it may seem to our psychologically more enlightened century), British naval tradition dictated that a captain have virtually no social contact with anyone down the chain of command. He usually dined alone and met with his officers primarily to discuss ship's business and to converse in the most formal and "correct" manner.

Now Fitzroy, when he set sail with Darwin, was only 26 years old. He knew the psychological toll that prolonged lack of human contact could take from captains. The *Beagle*'s previous skipper had broken down and shot himself to death during the Southern Hemisphere winter of 1828, his third year away from home. Moreover, as Darwin himself affirmed in a letter to his sister, Fitzroy was worried about "his hereditary predisposition" to mental derangement. His illustrious uncle, the Viscount Castlereagh (suppressor of the Irish rebellion of 1798 and Foreign Secretary during the defeat of Napoleon), had slit his own throat in 1822. In fact, Fitzroy did break down and temporarily relinquish his command during the *Beagle*'s voyage—while Darwin was laid up with illness in Valparaiso.

Since Fitzroy had so little social contact with any of the ship's official personnel, he could gain it only by taking along a "supernumerary" passenger by his own arrangement. But the Admiralty frowned upon private passengers, even captains' wives; a gentleman companion brought for no other stated purpose would never do. Fitzroy had taken other supernumeraries aboard—a draftsman and an instrument-maker among others—but neither could serve as a companion because they were not of the right social class. Fitzroy was an aristocrat, and he traced his ancestry directly to King Charles II. Only a gentleman could share his meals, and a gentleman Darwin surely was.

But how could Fitzroy entice a gentleman to accompany him on a voyage of five years' duration? Only by providing an opportunity for some justifying activity that could not be pursued elsewhere. And what else but natural history?—even though the *Beagle* had an official naturalist. Hence, Fitzroy advertised among his aristocratic friends for a gentleman naturalist. It was, as Burstyn argues, "A polite fiction to explain his guest's presence and an activity attractive enough to lure a gentleman on board for a long voyage." Darwin's sponsor, J. S. Henslow, understood perfectly. He wrote to Darwin: "Capt. F. wants a man (I understand) more as a companion than a mere collector." Darwin and Fitzroy met, they hit it off, and the pact was sealed. Darwin sailed as Fitzroy's companion, primarily to share his table at mealtime for every shipboard dinner during five long years. Fitzroy, in addition, was an ambitious young man. He wished to make his mark by setting a new standard for excellence in exploratory voyages. ("The object of the expedition," Darwin wrote, "was to complete the survey of Patagonia and Tierra del Fuego . . . to survey the shores of Chile, Peru, and of some islands in the Pacific—and to carry a chain of chronometrical measurements round the world.") By augmenting the official crew with technicians and engineers brought at his own expense, Fitzroy used his wealth and prestige to reach his goal. A "supernumerary" naturalist meshed well with Fitzroy's scheme to beef up the *Beagle*'s scientific mettle.

Poor McKormick's fate was sealed. Initially, he and Darwin cooperated, but their ways inevitably parted. Darwin had all the advantages. He had the captain's ear. He had a servant. At ports of call, he had the money to move ashore and hire native collectors, while McKormick was bound to the ship and his official duties. Darwin's private efforts began to outstrip McKormick's official collections, and McKormick, in disgust, decided to go home. In April 1832, at Rio de Janeiro, he was "invalided out" and sent home to England aboard H.M.S. *Tyne.* Darwin understood the euphemism and wrote to his sister of McKormick's "being invalided, i.e. being disagreeable to the Captain. . . . He is no loss."

Darwin did not care for McKormick's brand of science. He wrote to Henslow in May 1832: "He was a philosopher of rather antient [*sic*] date; at St. Jago by his own account he made general remarks during the first fortnight and collected particular facts during the last." In fact, Darwin didn't seem to care for McKormick at all. "My friend the doctor is an ass, but we jog on very amicably; at present he is in great tribulation, whether his cabin shall be painted french gray or dead white—I hear little except this subject from him."

If nothing else, this story illustrates the importance of social class as a consideration in the history of science. How different would the science of biology be today if Darwin had been the offspring of a tradesman and not the son of a very wealthy physician. Darwin's personal riches gave him the freedom to pursue research without encumbrance. Since his various illnesses often permitted only two to three hours of fruitful work per day, any need to make an honest living would probably have shut him off from research entirely. We now learn that Darwin's social standing also played a crucial role at a turning point in his career. Fitzroy was more interested in his mealtime companion's social graces than his competence in natural history.

Might something deeper be hidden in the unrecorded mealtime conversations of Darwin and Fitzroy? Scientists have a strong bias for attributing creative insights to the constraints of empirical evidence. Hence, tortoises and finches have always received the nod as primary agents in the transformation of Darwin's world view, for he joined the *Beagle* as a naïvely pious student for the ministry, but opened his first notebook on the transmutation of species less than a year after his return. I suggest that Fitzroy himself might have been an even more important catalyst.

Darwin and Fitzroy maintained a tense relationship at best. Only the severe constraints of gentlemanly cordiality and pre-Victorian suppression of emotion kept the two men on decent terms with each other. Fitzroy was a martinet and an ardent Tory. Darwin was an equally committed Whig. Darwin scrupulously avoided any discussion with Fitzroy of the great Reform Bill then pending in Parliament. But slavery brought them into open conflict. One evening, Fitzroy told Darwin that he had witnessed proof of slavery's benevolence. One of Brazil's largest slaveholders had assembled his captives and asked them whether they wished to be freed. Unanimously, they had responded "no." When Darwin had the temerity to wonder what a response made in the owner's presence was worth, Fitzroy exploded and informed Darwin that anyone who doubted his word was not fit to eat with him. Darwin moved out and joined the mates, but Fitzroy backed down and sent a formal apology a few days later.

We know that Darwin bristled in the face of Fitzroy's strong opinions. But he was Fitzroy's guest and, in one peculiar sense, his subordinate, for a captain at sea was an absolute and unquestioned tyrant in Fitzroy's time. Darwin could not express his dissent. For five long years, one of the most brilliant men in recorded history kept his peace. Late in life, Darwin recalled in his autobiography that "the difficulty of living on good terms with a Captain of a Man-of-War is much increased by its being almost mutinous to answer him as one would answer anyone else; and by the awe in which he is held —or was held in my time, by all on board."

Now Tory politics was not Fitzroy's only ideological passion. The other was religion. Fitzroy had some moments of doubt about the Bible's literal truth, but he tended to view Moses as an accurate historian and geologist and even spent considerable time trying to calculate the dimensions of Noah's Ark. Fitzroy's *idée fixe,* at least in later life, was the "argument from design," the belief that God's benevolence (indeed his very existence) can be inferred from the perfection of organic structure. Darwin, on the other hand, accepted the idea of excellent design but proposed a natural explanation that could not have been more contrary to Fitzroy's conviction. Darwin developed an evolutionary theory based on chance variation and natural selection imposed by an external environment: a rigidly materialistic (and basically atheistic) version of evolution (see essay 1). Many other evolutionary theories of the nineteenth century were far more congenial to Fitzroy's type of Christianity. Religious leaders, for example, had far less trouble with common proposals for innate perfecting tendencies than with Darwin's uncompromisingly mechanical view.

Was Darwin led to his philosophical outlook partly as a response to Fitzroy's dogmatic insistence upon the argument from design? We have no evidence that Darwin, aboard the *Beagle,* was anything but a good Christian. The doubts and rejection came later. Midway through the voyage, he wrote to a friend: "I often conjecture what will become of me; my wishes certainly would make me a country clergyman." And he even coauthored with Fitzroy an appeal for the support of Pacific missionary work entitled, "The Moral State of Tahiti." But the seeds of doubt must have been sown in quiet hours of contemplation aboard the *Beagle.* And think of Darwin's position on board—dining every day for five years with an authoritarian captain whom he could not rebuke, whose politics and bearing stood against all his beliefs, and whom, basically, he did not like. Who knows what "silent alchemy" might have worked upon Darwin's brain during five years of insistent harangue. Fitzroy may well have been far more important than finches, at least for inspiring the materialistic and antitheistic tone of Darwin's philosophy and evolutionary theory.

Fitzroy, at least, blamed himself as his mind became unhinged in later life. He began to see himself as the unwitting agent of Darwin's heresy (indeed, I am suggesting that this may be true in a more literal sense than Fitzroy ever imagined). He developed a burning desire to expiate his guilt and to reassert the Bible's supremacy. At the famous British Association Meeting of 1860 (where Huxley creamed Bishop "Soapy Sam" Wilberforce), the unbalanced Fitzroy stalked about, holding a Bible above his head and shouting, "The Book, The Book." Five years later, he slit his throat.

3 Darwin's Dilemma: The Odyssey of Evolution

THE EXEGESIS OF evolution as a concept has occupied the lifetimes of a thousand scientists. In this essay, I present something almost laughably narrow in comparison—an exegesis of the word itself. I shall trace how organic change came to be called *evolution.* The tale is complex and fascinating as a purely antiquarian exercise in etymological detection. But more is at stake, for a past usage of this word has contributed to the most common, current misunderstanding among laymen of what scientists mean by evolution.

To begin with a paradox: Darwin, Lamarck, and Haeckel—the greatest nineteenth-century evolutionists of England, France, and Germany, respectively—did not use the word evolution in the original editions of their great works. Darwin spoke of "descent with modification," Lamarck of "transformisme." Haeckel preferred "Transmutations-Theorie" or "Descendenz-Theorie." Why did they not use "evolution" and how did their story of organic change acquire its present name?

Darwin shunned evolution as a description of his theory for two reasons. In his day, first of all, evolution already had a technical meaning in biology. In fact, it described a theory of embryology that could not be reconciled with Darwin's views of organic development.

In 1744, the German biologist Albrecht von Haller had coined the term *evolution* to describe the theory that embryos grew from preformed homunculi enclosed in the egg or sperm (and that, fantastic as it may seem today, all future generations had been created in the ovaries of Eve or testes of Adam, enclosed like Russian dolls, one within the next—a homunculus in each of Eve's ova, a tinier homunculus in each ovum of the homunculus, and so on). This theory of evolution (or preformation) was opposed by the epigeneticists who believed that the complexity of adult shape arose from an originally formless egg (see essay 25 for a fuller account of this debate). Haller chose his term carefully, for the Latin *evolvere* means "to unroll"; indeed, the tiny homunculus unfolded from its originally cramped quarters and simply increased in size during its embryonic development.

Yet Haller's embryological evolution seemed to preclude Darwin's descent with modification. If the entire history of the human race were prepackaged into Eve's ovaries, how could natural selection (or any other force for that matter) alter the preordained course of our sojourn on earth?

Our mystery seems only to deepen. How could Haller's term be transformed into a nearly opposite meaning? This became possible only because Haller's theory was in its death throes by 1859; with its demise, the term that Haller had used became available for other purposes.

"Evolution" as a description of Darwin's "descent with modification" was not borrowed from a previous technical meaning; it was, rather, expropriated from the vernacular. Evolution, in Darwin's day, had become a common English word with a meaning quite different from Haller's technical sense. The *Oxford English Dictionary* traces it to a 1647 poem of H. More: "Evolution of outward forms spread in the world's vast spright [spirit]." But this was "unrolling" in a sense very different from Haller's. It implied "the appearance in orderly succession of a long train of events," and more important, it embodied a *concept of progressive development*—an orderly unfolding from simple to complex. The *O.E.D.* continues, "The process of developing from a rudimentary to a mature or complete state." Thus evolution, in the vernacular, was firmly tied to a concept of progress.

Darwin did use evolve in this vernacular sense—in fact it is the very last word of his book.

> There is grandeur in this view of life, with its several powers, having been originally breathed into a few forms or into one; and that, whilst this planet has gone cycling on according to the fixed law of gravity, from so simple a beginning endless forms most beautiful and most wonderful have been, and are being evolved.

Darwin chose it for this passage because he wanted to contrast the flux of organic development with the fixity of such physical laws as gravitation. But it was a word he used very rarely indeed, for Darwin explicitly rejected the common equation of what we now call evolution with any notion of progress.

In a famous epigram, Darwin reminded himself never to say "higher" or "lower" in describing the structure of organisms—for if an amoeba is as well adapted to its environment as we are to ours, who is to say that we are higher creatures? Thus Darwin shunned evolution as a description for his descent with modification, both because its technical meaning contrasted with his beliefs and because he was uncomfortable with the notion of inevitable progress inherent in its vernacular meaning.

Evolution entered the English language as a synonym for "descent with modification" through the propaganda of Herbert Spencer, that indefatigable Victorian pundit of nearly everything. Evolution, to Spencer, was the overarching law of all development. And, to a smug Victorian, what principle other than progress could rule the developmental processes of the universe? Thus, Spencer defined the universal law in his *First Principles* of 1862: "Evolution is an integration of matter and concomitant dissipation of motion; during which the matter passes from an indefinite, incoherent homogeneity to a definite coherent heterogeneity."

Two other aspects of Spencer's work contributed to the establishment of evolution in its present meaning: First, in writing his very popular *Principles of Biology* (1864–67), Spencer constantly used "evolution" as a description of organic change. Second, he did not view progress as an intrinsic capacity of matter, but as a result of "cooperation" between internal and external (environmental) forces. This view fit nicely with most nineteenth-century concepts of organic evolution, for Victorian scientists easily equated organic change with organic progress. Thus evolution was available when many scientists felt a need for a term more succinct than Darwin's descent with modification. And since most evolutionists saw organic change as a process directed toward increasing complexity (that is, to us), their appropriation of Spencer's general term did no violence to his definition.

Ironically, however, the father of evolutionary theory stood almost alone in insisting that organic change led only to increasing adaptation between organisms and their own environment and not to an abstract ideal of progress defined by structural complexity or increasing heterogeneity—never say higher or lower. Had we heeded Darwin's warning, we would have been spared much of the confusion and misunderstanding that exists between scientists and laymen today. For Darwin's view has triumphed among scientists who long ago abandoned the concept of necessary links between evolution and progress as the worst kind of anthropocentric bias. Yet most laymen still equate evolution with progress and define human evolution not simply as change, but as increasing intelligence, increasing height, or some other measure of assumed improvement.

In what may well be the most widespread antievolutionary document of modern times, the Jehovah's Witnesses' pamphlet "Did Man Get Here by Evolution or by Creation?" proclaims: "Evolution, in very simple terms, means that life progressed from one-celled organisms to its highest state, the human being, by means of a series of biological changes taking place over millions of years. . . . Mere change within a basic type of living thing is not to be regarded as evolution."

This fallacious equation of organic evolution with progress continues to have unfortunate consequences. Historically, it engendered the abuses of Social Darwinism (which Darwin himself held in such suspicion). This discredited theory ranked human groups and cultures according to their assumed level of evolutionary attainment, with (not surprisingly) white Europeans at the top and people dwelling in their conquered colonies at the bottom. Today, it remains a primary component of our global arrogance, our belief in dominion over, rather than fellowship with, more than a million other species that inhabit our planet. The moving finger has written, of course, and nothing can be done; yet I am rather sorry that scientists contributed to a fundamental misunderstanding by selecting a vernacular word meaning progress as a name for Darwin's less euphonious but more accurate "descent with modification."

4 Darwin's Untimely Burial

IN ONE OF THE numerous movie versions of *A Christmas Carol*, Ebenezer Scrooge encounters a dignified gentleman sitting on a landing, as he mounts the steps to visit his dying partner, Jacob Marley. "Are you the doctor?" Scrooge inquires. "No," replies the man, "I'm the undertaker; ours is a very competitive business." The cutthroat world of intellectuals must rank a close second, and few events attract more notice than a proclamation that popular ideas have died. Darwin's theory of natural selection has been a perennial candidate for burial. Tom Bethell held the most recent wake in a piece called "Darwin's Mistake" (*Harper's*, February 1976): "Darwin's theory, I believe, is on the verge of collapse. . . . Natural selection was quietly abandoned, even by his most ardent supporters, some years ago." News to me, and I, although I wear the Darwinian label with some pride, am not among the most ardent defenders of natural selection. I recall Mark Twain's famous response to a premature obituary: "The reports of my death are greatly exaggerated."

Bethell's argument has a curious ring for most practicing scientists. We are always ready to watch a theory fall under the impact of new data, but we do not expect a great and influential theory to collapse from a logical error in its formulation. Virtually every empirical scientist has a touch of the Philistine. Scientists tend to ignore academic philosophy as an empty pursuit. Surely, any intelligent person can think straight by intuition. Yet Bethell cites no data at all in sealing the coffin of natural selection, only an error in Darwin's reasoning: "Darwin made a mistake sufficiently serious to undermine his theory. And that mistake has only recently been recognized as such. . . . At one point in his argument, Darwin was misled."

Although I will try to refute Bethell, I also deplore the unwillingness of scientists to explore seriously the logical structure of arguments. Much of what passes for evolutionary theory is as vacuous as Bethell claims. Many great theories are held together by chains of dubious metaphor and analogy. Bethell has correctly identified the hogwash surrounding evolutionary theory. But we differ in one fundamental way: for Bethell, Darwinian theory is rotten to the core; I find a pearl of great price at the center.

Natural selection is the central concept of Darwinian theory—the fittest survive and spread their favored traits through populations. Natural selection is defined by Spencer's phrase "survival of the fittest," but what does this famous bit of jargon really mean? Who are the fittest? And how is "fitness" defined? We often read that fitness involves no more than "differential reproductive success"—the production of more surviving offspring than other competing members of the population. Whoa! cries Bethell, as many others have before him. This formulation defines fitness in terms of survival only. The crucial phrase of natural selection means no more than "the survival of those who survive"—a vacuous tautology. (A tautology is a phrase—like "my father is a man" —containing no information in the predicate ("a man") not inherent in the subject ("my father"). Tautologies are fine as definitions, but not as testable scientific statements—there can be nothing to test in a statement true by definition.)

But how could Darwin have made such a monumental, two-bit mistake? Even his severest critics have never accused him of crass stupidity. Obviously, Darwin must have tried to define fitness differently—to find a criterion for fitness independent of mere survival. Darwin did propose an independent criterion, but Bethell argues quite correctly that he relied upon analogy to establish it, a dangerous and slippery strategy. One might think that the first chapter of such a revolutionary book as *Origin of Species* would deal with cosmic questions and general concerns. It doesn't. It's about pigeons. Darwin devotes most of his first forty pages to "artificial selection" of favored traits by animal breeders. For here an independent criterion surely operates. The pigeon fancier knows what he wants. The fittest are not defined by their survival. They are, rather, allowed to survive because they possess desired traits.

The principle of natural selection depends upon the validity of an analogy with artificial selection. We must be able, like the pigeon fancier, to identify the fittest beforehand, not only by their subsequent survival. But nature is not an animal breeder; no preordained purpose regulates the history of life. In nature, any traits possessed by survivors must be counted as "more evolved"; in artificial selection, "superior" traits are defined before breeding even begins. Later evolutionists, Bethell argues, recognized the failure of Darwin's analogy and redefined "fitness' as mere survival. But they did not realize that they had undermined the logical structure of Darwin's central postulate. Nature provides no independent criterion of fitness; thus, natural selection is tautological.

Bethel then moves to two important corollaries of his major argument. First, if fitness only means survival, then how can natural selection be a "creative" force, as Darwinians insist. Natural selection can only tell us how "a given type of animal became more numerous"; it cannot explain "how one type of animal gradually changed into another." Secondly, why were Darwin and other eminent Victorians so sure that mindless nature could be compared with conscious selection by breeders. Bethell argues that the cultural climate of triumphant industrial capitalism had defined any change as inherently progressive. Mere survival in nature could only be for the good:"It is beginning to look as though what Darwin really discovered was nothing more than the Victorian propensity to believe in progress."

I believe that Darwin was right and that Bethell and his colleagues are mistaken: criteria of fitness independent of survival can be applied to nature and have been used consistently by evolutionists. But let me first admit that Bethell's criticism applies to much of the technical literature in evolutionary theory, especially to the abstract mathematical treatments that consider evolution only as an alteration in numbers, not as a change in quality. These studies do assess fitness only in terms of differential survival. What else can be done with abstract models that trace the relative successes of hypothetical genes A and B in populations that exist only on computer tape? Nature, however, is not limited by the calculations of theoretical geneticists. In nature, A's "superiority" over B will be *expressed* as differential survival, but it is not *defined* by it—or, at least, it better not be so defined, lest Bethell et al. triumph and Darwin surrender.

My defense of Darwin is neither startling, novel, nor profound. I merely assert that Darwin was justified in analogizing natural selection with animal breeding. In artificial selection, a breeder's desire represents a "change of environment" for a population. In this new environment, certain traits are superior a priori; (they survive and spread by our breeder's choice, but this is a *result* of their fitness, not a definition of it). In nature, Darwinian evolution is also a response to changing environments. Now, the key point: certain morphological, physiological, and behavioral traits should be superior a priori as designs for living in new environments. These traits confer fitness by an engineer's criterion of good design, not by the empirical fact of their survival and spread. It got colder before the woolly mammoth evolved its shaggy coat.

Why does this issue agitate evolutionists so much? OK, Darwin was right: superior design in changed environments is an independent criterion of fitness. So what? Did anyone ever seriously propose that the poorly designed shall triumph? Yes, in fact, many did. In Darwin's day, many rival evolutionary theories asserted that the fittest (best designed) must perish. One popular notion—the theory of racial life cycles—was championed by a former inhabitant of the office I now occupy, the great American paleontologist Alpheus Hyatt. Hyatt claimed that evolutionary lineages, like individuals, had cycles of youth, maturity, old age, and death (extinction). Decline and extinction are programmed into the history of species. As maturity yields to old age, the best-designed individuals die and the hobbled, deformed creatures of phyletic senility take over. Another anti-Darwinian notion, the theory of orthogenesis, held that certain trends, once initiated, could not be halted, even though they must lead to extinction caused by increasingly inferior design. Many nineteenth-century evolutionists (perhaps a majority) held that Irish elks became extinct because they could not halt their evolutionary increase in antler size (see essay 9); thus, they died—caught in trees or bowed (literally) in the mire. Likewise, the demise of saber-toothed "tigers" was often attributed to canine teeth grown so long that the poor cats couldn't open their jaws wide enough to use them.

Thus, it is not true, as Bethell claims, that any traits possessed by survivors must be designated as fitter. "Survival of the fittest" is not a tautology. It is also not the only imaginable or reasonable reading of the evolutionary record. It is testable. It had rivals that failed under the weight of contrary evidence and changing attitudes about the nature of life. It has rivals that may succeed, at least in limiting its scope.

If I am right, how can Bethell claim, "Darwin, I suggest, is in the process of being discarded, but perhaps in deference to the venerable old gentleman, resting comfortably in Westminster Abbey next to Sir Isaac Newton, it is being done as discreetly and gently as possible with a minimum of publicity." I'm afraid I must say that Bethell has not been quite fair in his report of prevailing opinion. He cites the gadflies C. H. Waddington and H. J. Muller as though they epitomized a consensus. He never mentions the leading selectionists of our present generation—E. O. Wilson or D. Janzen, for example. And he quotes the architects of neo-Darwinism—Dobzhansky, Simpson, Mayr, and J. Huxley—only to ridicule their metaphors on the "creativity" of natural selection. (I am not claiming that Darwinism should be cherished because it is still popular; I am enough of a gadfly to believe that uncriticized consensus is a sure sign of impending trouble. I merely report that, for better or for worse, Darwinism is alive and thriving, despite Bethell's obituary.)

But why was natural selection compared to a composer by Dobzhansky; to a poet by Simpson; to a sculptor by Mayr; and to, of all people, Mr. Shakespeare by Julian Huxley? I won't defend the choice of metaphors, but I will uphold the intent, namely, to illustrate the essence of Darwinism—the creativity of natural selection. Natural selection has a place in all anti-Darwinian theories that I know. It is cast in a negative role as an executioner, a headsman for the unfit (while the fit arise by such non-Darwinian mechanisms as the inheritance of acquired characters or direct induction of favorable variation by the environment). The essence of Darwinism lies in its claim that natural selection creates the fit. Variation is ubiquitous and random in direction. It suppiies the raw material only. Natural selection directs the course of evolutionary change. It preserves favorable variants and builds fitness gradually. In fact, since artists fashion their creations from the raw material of notes, words, and stone, the metaphors do not strike me as inappropriate. Since Bethell does not accept a criterion of fitness independent of mere survival, he can hardly grant a creative role to natural selection.

According to Bethell, Darwin's concept of natural selection as a creative force can be no more than an illusion encouraged by the social and political climate of his times. In the throes of Victorian optimism in imperial Britain, change seemed to be inherently progressive; why not equate survival in nature with increasing fitness in the nontautological sense of improved design.

I am a strong advocate of the general argument that "truth" as preached by scientists often turns out to be no more than prejudice inspired by prevailing social and political beliefs. I have devoted several essays to this theme because I believe that it helps to "demystify" the practice of science by showing its similarity to all creative human activity. But the truth of a general argument does not validate any specific application, and I maintain that Bethell's application is badly misinformed.

Darwin did two very separate things: he convinced the scientific world that evolution had occurred and he proposed the theory of natural selection as its mechanism. I am quite willing to admit that the common equation of evolution with progress made Darwin's first claim more palatable to his contemporaries. But Darwin failed in his second quest during his own lifetime. The theory of natural selection did not triumph until the 1940s. Its Victorian unpopularity, in my view, lay primarily in its denial of general progress as inherent in the workings of evolution. Natural selection is a theory of *local* adaptation to changing environments. It proposes no perfecting principles, no guarantee of general improvement; in short, no reason for general approbation in a political climate favoring innate progress in nature.

Darwin's independent criterion of fitness is, indeed, "improved design," but not "improved" in the cosmic sense that contemporary Britain favored. To Darwin, improved meant only "better designed for an immediate, local environment." Local environments change constantly: they get colder or hotter, wetter or drier, more grassy or more forested. Evolution by natural selection is no more than a tracking of these changing environments by differential preservation of organisms better designed to live in them: hair on a mammoth is not progressive in any cosmic sense. Natural selection can produce a trend that tempts us to think of more general progress—increase in brain size does characterize the evolution of group after group of mammals (see essay 23). But big brains have their uses in local environments; they do not mark intrinsic trends to higher states. And Darwin delighted in showing that local adaptation often produced "degeneration" in design—anatomical simplification in parasites, for example.

If natural selection is not a doctrine of progress, then its popularity cannot reflect the politics that Bethell invokes. If the theory of natural selection contains an independent criterion of fitness, then it is not tautological. I maintain, perhaps naïvely, that its current, unabated popularity must have something to do with its success in explaining the admittedly imperfect information we now possess about evolution. I rather suspect that we'll have Charles Darwin to kick around for some time.

2 Human Evolution

5 | A Matter of Degree

IN *Alexander's Feast,* John Dryden describes his hero, besotted after dinner, retelling the tales of his martial glory:

> The King grew vain;
> Fought all his battles o'er again;
> And thrice he routed all his foes,
> and thrice he slew the slain.

One hundred and fifty years later, Thomas Henry Huxley invoked the same image in declining to pursue further the decisive victory he had won over Richard Owen in the great hippocampus debate: "Life is too short to occupy oneself with the slaying of the slain more than once."

Owen had sought to establish our uniqueness by arguing that a small convolution of the human brain, the hippocampus minor, was absent in chimps and gorillas (and all other creatures), but present in *Homo sapiens* alone. Huxley, who had been dissecting primates while preparing his seminal work, *Evidence as to Man's Place in Nature,* showed conclusively that all apes had a hippocampus, and that any discontinuity in the structure of primate brains lay between prosimians (lemurs and tarsiers) and all other primates (including humans), not between man and the great apes. Yet for a month, in April, 1861, all England watched as her two greatest anatomists waged war over a little bump on the brain. *Punch* laughed and versified; and Charles Kingsley wrote at length of the "hippopotamus major" in his children's classic of 1863, *The Water Babies.* If a water baby had ever been found, he commented, "they would have put it into spirits, or into the *Illustrated News,* or perhaps cut it into two halves, poor dear little thing, and sent one to Professor Owen, and one to Professor Huxley, to see what they could each say about it."

The Western world has yet to make its peace with Darwin and the implications of evolutionary theory. The hippocampus debate merely illustrates, in light relief, the greatest impediment to this reconciliation—our unwillingness to accept continuity between ourselves and nature, our ardent search for a criterion to assert our uniqueness. Again and again, the great naturalists have enunciated general theories of nature and made singular exceptions for humans. Charles Lyell (see essay 18) envisioned a world in steady-state: no change through time in the complexity of life, with all organic designs present from the first. Yet man alone was created but a geological instant ago—a quantum jump in the moral sphere imposed upon the constancy of mere anatomical design. And Alfred Russel Wallace, an ardent selectionist who far out-Darwined Darwin in his rigid insistence on natural selection as the sole directing force for evolutionary change, made his only exception for the human brain (and turned to spiritualism late in his life).

Darwin himself, although he accepted strict continuity, was reluctant to expose his heresy. In the first edition of the *Origin of Species* (1859), he wrote only that "light will be thrown on the origin of man and his history." Later editions added the intensifier "much" before the sentence. Only in 1871 did he gather the courage to publish *The Descent of Man* (see essay 1).

Chimps and gorillas have long been the battleground of our search for uniqueness; for if we could establish an unambiguous distinction—of kind rather than of degree—between ourselves and our closest relatives, we might gain the justification long sought for our cosmic arrogance. The battle shifted long ago from a simple debate about evolution: educated people now accept the evolutionary continuity between humans and apes. But we are so tied to our philosophical and religious heritage that we still seek a criterion for strict division between our abilities and those of chimpanzees. For, as the psalmist sang: "What is man, that thou art mindful of him? . . . For thou has made him a little lower than the angels, and hast crowned him with glory and honor." Many criteria have been tried, and one by one they have failed. The only honest alternative is to admit the strict continuity in kind between ourselves and chimpanzees. And what do we lose thereby? Only an antiquated concept of soul to gain a more humble, even exalting vision of our oneness with nature. I propose to examine three criteria for distinction and to argue that, on all accounts, we are more nearly akin to the chimpanzee than even Huxley dared to think.

1. Morphological uniqueness in the Owenian tradition. Huxley permanently dimmed the ardor of those seeking an anatomical discontinuity between humans and apes. Still, the search has continued in some quarters. The differences between adult chimps and people are not trifling, but they do not arise from any difference in kind. Part by part, order by order, we are the same; only relative sizes and rates of growth differ. With the painstaking attention to detail so characteristic of German anatomical research, Prof. D. Starck and his colleagues have recently concluded that differences between the skulls of humans and chimps are quantitative only.

2. Conceptual uniqueness. Few scientists have strongly pushed the anatomical argument since Owen's debacle. Instead, the defenders of human uniqueness have posited an unbridgeable chasm between the mental abilities of humans and chimps. To illustrate the gap, they have sought an unambiguous criterion of distinction. An earlier generation cited use of tools, but clever chimps employ all sorts of artifacts to reach inaccessible bananas or release imprisoned mates.

More recent claims have centered on language and conceptualization, the last bastion for potential differences in kind. Early experiments on teaching chimps to talk were notably unsuccessful—a few grunts and a trifling vocabulary. Some concluded that the failure must reflect a deficiency in cerebral organization, but the explanation seems simpler and far less profound (although by no means unimportant for what it implies about the linguistic capabilities of chimps in natural conditions): the vocal cords of chimpanzees are constructed in such a way that large repertories of articulated sounds cannot be produced. If we could only discover a different way of communicating with them, we might find that chimps are much smarter than we think.

By now, all readers of newspapers and watchers of television have learned of the striking initial successes of another way—communicating with chimps via sign language of the deaf and dumb. When Lana, star pupil of the Yerkes Laboratory, began to ask for the names of objects she had not previously seen, can we any longer deny to chimps the capacity to conceptualize and to abstract? This is no mere Pavlovian conditioning. In February, 1975, R. A. and B. T. Gardner reported their first results on two baby chimpanzees raised with sign language from the day of their birth. (Washoe, their previous subject, was not exposed to sign language until she was a year old. After six months of training, her vocabulary consisted of only two signs.) Both baby chimps began to make recognizable signs in their third month. One, Moja, had a four-word vocabulary in her thirteenth week: come-gimme, go, more, and drink. Their current progress is no slower than that of a human child (we generally wait for words and do not realize that our babies signal us in other ways long before they speak). Of course, I do not believe that our mental differences with chimps are merely a question of nurturing. I have no doubt that the progress of these baby chimps will slow down relative to the growing achievements of human babies. The next president of our country will not belong to another species. Still, the Gardners' work is a striking demonstration of how we have underestimated our closest biological relatives.

3 . Overall genetic differences. Even if we admit that no single feature or ability completely separates humans and chimps, at least we might be able to affirm that the overall genetic differences between us are tolerably great. After all, the two species look very different and do very different things under natural conditions. (For all the quasi-linguistic

capacity shown by chimps in the laboratory, we have no evidence of rich conceptual communication in the wild.) But Mary-Claire King and A. C. Wilson have recently published an account of genetic differences between the two species (*Science,* April 11, 1975), and the results may well upset a prior prejudice still carried, I suspect, by most of us. In short, using all the biochemical techniques now available and surveying as many proteins as possible, the overall genetic differences are remarkably small.

When two species scarcely differ in morphology but function as separate and reproductively isolated populations in nature, evolutionary biologists speak of "sibling species." Sibling species generally display far fewer genetic differences than pairs of species placed in the same genus but clearly different in morphology ("congeneric species"). Now chimps and humans are obviously not sibling species; we are not even congeneric species by conventional taxonomic practice (chimps belong to the genus *Pan;* we are *Homo sapiens*). But King and Wilson have shown that the overall genetic distance between humans and chimps is less than the average for sibling species and far less than in any tested pair of congeneric species.

A fine paradox, for although I have argued strongly that our distinctions are matters of degree only, we are still very different animals. If the overall genetic distance is so small, then what has caused such a divergence in form and behavior? Under the atomistic notion that each organic trait is controlled by a single gene, we cannot reconcile our anatomical dissimilarities with King and Wilson's findings, for many differences in form and function would have to reflect many differences in genes.

The answer must be that certain kinds of genes have far-reaching effects—they must influence the entire organism, not just single traits. A few changes in these key genes might produce a great divergence between two species without much overall genetic differentiation. King and Wilson therefore seek to resolve the paradox by attributing our differences from chimps primarily to mutations of the regulatory system.

Liver cells and brain cells have all the same chromosomes and all the same genes. Their profound difference does not arise from genetic constitution, but from alternate paths of development. During development, different genes must be turned on and off at different times in order to achieve such disparate results from the same genetic system. In fact, the whole mysterious process of embryology must be regulated by exquisite timing in the action of genes. To differentiate a hand from a homogeneous limb bud, for example, cells must proliferate in some areas (destined to be fingers) and die in others (the spaces between them).

Much of the genetic system must be devoted to setting the timing of these events—to turning genes on and off—rather than to the determination of specific traits. We refer to genes that control the timing of developmental events as the regulatory system. Clearly, change in a single regulatory gene can have profound effects upon the entire organism. Delay or accelerate a key event in embryology and the whole course of future development may be changed. King and Wilson therefore suppose that the primary genetic differences between humans and chimps lie in this all-important regulatory system.

This is a reasonable (even necessary) hypothesis. But do we know anything about the nature of this regulatory difference? We cannot now identify the specific genes involved; hence, King and Wilson express no opinion. "Most important for the future study of human evolution," they write, "would be the demonstration of differences between apes and humans in the timing of gene expression during development." But I believe that we do know the basis of this change in timing. As I argue in essay 7, *Homo sapiens* is basically a neotenic species; we have evolved from apelike ancestors by a general retardation in developmental rate.We should look for regulatory changes that slow down the ontogenetic trends we share with all primates and allow us to retain juvenile growth tendencies and proportions.

The very small genetic distance between humans and chimps might tempt us to try the most potentially interesting and ethically unacceptable scientific experiment I can imagine—to hybridize our two species and simply to ask the offspring what it is like to be, at least in part, a chimpanzee. This interbreeding may well be possible—so small are the genetic distances that separate us. But, lest we fear the rise of a race comparable to the heroes in Planet of the Apes, I hasten to add that the hybrids would almost certainly be sterile—like a mule, and for the same reason. The genetic differences between humans and chimps are minor, but they include at least ten large inversions and translocations. An inversion is, literally, the turning around of a chromosomal segment. Each hybrid cell would have a set of chimp and a corresponding set of human chromosomes. Egg and sperm cells are made by a process called meiosis, or reduction division. In meiosis, each chromosome must pair (lie side by side) with its counterpart before cell division, so that corresponding genes can match up one to one: that is, each chimp chromosome must pair with its human counterpart. But if a piece of human chromosome is inverted relative to its counterpart in chimps, then gene-by-gene pairing cannot occur without elaborate looping and twisting that usually precludes successful cell division.

The temptations are great, but I trust that this pairing will remain on the index of forbidden experiments. The temptation, in any case, will surely diminish as we discover how to talk with our closest relatives. I am beginning to suspect that we will learn everything we want to know directly from the chimps themselves.

6 | Bushes and Ladders in Human Evolution

MY FIRST TEACHER of paleontology was almost as old as some of the animals he discussed. He lectured from notes on yellow foolscap that he must have assembled during his own days in graduate school. The words changed not at all from year to year, but the paper got older and older. I sat in the first row, bathed in yellow dust, as the paper cracked and crumbled every time he turned a page.

It is a blessing that he never had to lecture on human evolution. New and significant prehuman fossils have been unearthed with such unrelenting frequency in recent years that the fate of any lecture notes can only be described with the watchword of a fundamentally irrational economy—planned obsolescence. Each year, when the topic comes up in my courses, I simply open my old folder and dump the contents into the nearest circular file. And here we go again.

A front-page headline in the *New York Times* for October 31, 1975, read: "Man traced 3.75 million years by fossils found in Tanzania." Dr. Mary Leakey, unsung hero of the famous clan, had discovered the jaws and teeth of at least eleven individuals in sediments located between two layers of fossil volcanic ash dated at 3.35 and 3.75 million years, respectively. (Mary Leakey, usually described only as Louis's widow, is a famous scientist with more impressive credentials than those of her flamboyant late husband. She also discovered several of the famous fossils usually attributed to Louis, including the "nutcracker man" of Olduvai, *Australopithecus boi-*

sei, their first important find.) Mary Leakey classified these fragments as the remains of creatures in our genus *Homo,* presumably of the East African species *Homo habilis,* first described by Louis Leakey.[2]

So what? In 1970, Harvard paleontologist Brian Patterson dated an East African jaw at 5.5 million years. True, he attributed the fragment to the genus *Australopithecus,* not to *Homo.* But *Australopithecus* has been widely regarded as the direct ancestor of *Homo.* While taxonomic convention requires the award of different names to stages of an evolving lineage, this custom should not obscure biological reality. If *H. habilis* is the direct descendant of *A. africanus* (and if the two species differ little in anatomical features), then the oldest "human" might as well be the oldest *Australopithecus,* not the oldest recipient of the arbitrary designation *Homo.* What, then, is so exciting about some jaws and teeth a million and a half years younger than the oldest *Australopithecus?*

I believe that Mary Leakey's find is the second most important discovery of the decade. To explain my excitement, I must provide some background in human paleontology and discuss a fundamental, but little appreciated, issue in evolutionary theory—the conflict between "ladders" and "bushes" as metaphors for evolutionary change. I want to argue that *Australopithecus,* as we know it, may not be the ancestor of *Homo;* and that, in any case, ladders do not represent the path of evolution. (By "ladders" I refer to the popular picture of evolution as a continuous sequence of ancestors and descendants.) Mary Leakey's jaws and teeth are the oldest "humans" we know.

---

2 | I wrote this essay in January, 1976. True to the admonition of my last paragraph, Mary Leakey's attribution of the Laetolil jaws to the genus *Homo* has been challenged by several colleagues. They assert no alternate hypothesis, but merely argue that jaws alone offer too little for a certain diagnosis. In any case, the primary assertion of this article remains valid—from our knowledge of African fossils, the genus *Homo* may be as old as the australopithecines. Moreover, we still have no firm evidence for any progressive change within any hominid species.

The metaphor of the ladder has controlled most thinking about human evolution. We have searched for a single, progressive sequence linking some apish ancestor with modern man by gradual and continuous transformation. The "missing link" might as well have been called the "missing rung." As the British biologist J. Z. Young recently wrote (1971) in his *Introduction to the Study of Man:* "Some interbreeding but varied population gradually changed until it reached the condition we recognize as that of *Homo sapiens.*"

Ironically, the metaphor of the ladder first denied a role in human evolution to the African australopithecines. *A. africanus* walked fully erect, but had a brain less than one-third the size of ours (see essay 22). When it was discovered in the 1920s, many evolutionists believed that all traits should change in concert within evolving lineages—the doctrine of the "harmonious transformation of the type." An erect, but small-brained ape could only represent an anomalous side branch destined for early extinction (the true intermediate, I assume, would have been a semierect, half-brained brute). But, as modern evolutionary theory developed during the 1930s, this objection to *Australopithecus* disappeared. Natural selection can work independently upon adaptive traits in evolutionary sequences, changing them at different times and rates. Frequently, a suite of characters undergoes a complete transformation before other characters change at all. Paleontologists refer to this potential independence of traits as "mosaic evolution."

Secured by mosaic evolution, *A. africanus* attained the exalted status of direct ancestor. Orthodoxy became a three-runged ladder: *A. africanus–H. erectus* (Java and Peking Man)–*H. sapiens.*

A small problem arose during the 1930s when another species of australopithecine was discovered—the so-called robust form, *A. robustus* (and later the more extreme "hyper-robust," *A. boisei,* found by Mary Leakey in the late 1950s). Anthropologists were forced to admit that two species of australopithecines lived contemporaneously and that the ladder contained at least one side branch. Still, the ancestral status of *A. africanus* was not challenged; it merely acquired a second and ultimately unsuccessful descendant, the small-brained, big-jawed robust lineage.

Then, in 1964, Louis Leakey and his colleagues began a radical reassessment of human evolution by naming a new species from East Africa, *Homo habilis.* They believed that *H. habilis* was a contemporary of the two australopithecine lineages; moreover, as the name implies, they regarded it as distinctly more human than either of its contemporaries. Bad news for the ladder: three coexisting lineages of prehumans! And a potential descendant *(H. habilis)* living at the same time as its presumed ancestors. Leakey proclaimed the obvious heresy: both lineages of australopithecines are side branches with no direct role in the evolution of *Homo sapiens.*

But *H. habilis,* as Leakey defined it, was controversial for two reasons. The conventional ladder could still be defended:

1. The fossils were scrappy and came from different places and times. Many anthropologists argued that Leakey's definition had mixed two different things, neither a new species: some older material properly assigned to *A. africanus,* and some younger fossils belonging to *H. erectus.*

2. The dating was insecure. Even if *H. habilis* represented a valid species, it might be younger than most or all of the known australopithecines. Orthodoxy could become a four-runged ladder: *A. africanus–H. habilis–H. erectus–H. sapiens.*

But, as a new consensus began to coalesce about the expanded ladder, Louis and Mary Leakey's son Richard reported the find of the decade in 1973 . He had unearthed a nearly complete skull with a cranial capacity near 800 cc, almost twice that of any*A. africanus* specimen. Moreover, and this is the crucial point, he dated the skull at between 2 and 3 million years, with a preference for something near the older figure—that is, older than most australopithecine fossils, and not far from the oldest, 5.5-million-year date. *H. habilis* was no longer a chimera of Louis's imagination. (Richard Leakey's specimen is often cautiously designated only by its field number, ER-1470. But whether or not we choose to use the name *Homo habilis,* it is surely a member of our genus, and it is just as surely a contemporary of *Australopithecus.*)

Mary Leakey has now extended the range of *H. habilis* back another million years (perhaps closer to 2 million years, if 1470 is closer to 2 than to 3 million years old, as many experts now believe). *H. habilis* is not the direct descendant of known *A. africanus;* the new finds are, in fact, older than almost all specimens of *A. africanus* (and the taxonomic status of all fragmentary specimens older than Mary Leakey's *H. habilis* is in doubt). Based on the fossils as we know them, *Homo* is as old as *Australopithecus.* (One can still argue that *Homo* evolved from an older, as yet undiscovered *Australopithecus.* But no evidence supports such a claim, and I could speculate with equal justice that *Australopithecus* evolved from an unknown *Homo.*)

Chicago anthropologist Charles Oxnard has dealt *Australopithecus* another blow from a different source. He studied the shoulder, pelvis, and foot of australopithecines, modern primates (great apes and some monkeys), and *Homo* with the rigorous techniques of multivariate analysis (the simultaneous statistical consideration of large numbers of measures). He concludes—though many anthropologists disagree—that the australopithecines were "uniquely different" from either apes or humans, and argues for "the removal of the different members of this relatively small-brained, curiously unique genus *Australopithecus* into one or more parallel side lines away from a direct link with man."

What has become of our ladder if we must recognize three coexisting lineages of hominids *(A. africanus,* the robust australopithecines, and *H. habilis),* none clearly derived from another? Moreover, none of the three display any evolutionary trends during their tenure on earth: none become brainier or more erect as they approach the present day.

At this point, I confess, I cringe, knowing full well what all the creationists who deluge me with letters must be thinking. "So Gould admits that we can trace no evolutionary ladder among early African hominids; species appear and later disappear, looking no different from their great-grandfathers. Sounds like special creation to me." (Although one might ask why the Lord saw fit to make so many kinds of hominids, and why some of his later productions, *H. erectus* in particular, look so much more human than the earlier models.) I suggest that the fault is not with evolution itself, but with a false picture of its operation that most of us hold—namely the ladder; which brings me to the subject of bushes.

I want to argue that the "sudden" appearance of species in the fossil record and our failure to note subsequent evolutionary change within them is the proper prediction of evolutionary theory as we understand it. Evolution usually proceeds by "speciation"—the splitting of one lineage from a parental stock—not by the slow and steady transformation of these large parental stocks. Repeated episodes of speciation produce a bush. Evolutionary "sequences" are not rungs on a ladder, but our retrospective reconstruction of a circuitous path running like a labyrinth, branch to branch, from the base of the bush to a lineage now surviving at its top.

How does speciation occur? This is a perennial hot topic in evolutionary theory, but most biologists would subscribe to the "allopatric theory" (the debate centers on the admissibility of other modes; nearly everyone agrees that allopatric speciation is the most common mode). *Allopatric* means "in another place." In the allopatric theory, popularized by Ernst Mayr, new species arise in *very small* populations that become isolated from their parental group at the *periphery* of the ancestral range. Speciation in these small isolates is *very rapid* by evolutionary standards—hundreds or thousands of years (a geological microsecond).

Major evolutionary change may occur in these small, isolated populations. Favorable genetic variation can quickly spread through them. Moreover, natural selection tends to be intense in geographically marginal areas where the species barely maintains a foothold. In large central populations, on the other hand, favorable variations spread very slowly, and most change is steadfastly resisted by the well-adapted population. Small changes occur to meet the requirements of slowly altering climates, but major genetic reorganizations almost always take place in the small, peripherally isolated populations that form new species.

If evolution almost always occurs by rapid speciation in small, peripheral isolates—rather than by slow change in large, central populations—then what should the fossil record look like? We are not likely to detect the event of speciation itself. It happens too fast, in too small a group, isolated too far from the ancestral range. We will first meet the new species as a fossil when it reinvades the ancestral range and becomes a large central population in its own right. During its recorded history in the fossil record, we should expect no major change; for we know it only as a successful, central population. It will participate in the process of organic change only when some of its peripheral isolates speciate to become new branches on the evolutionary bush. But it, itself, will appear "suddenly" in the fossil record and become extinct later with equal speed and little perceptible change in form.

The fossil hominids of Africa fully meet these expectations. We know about three coexisting branches of the human bush. I will be surprised if twice as many more are not discovered before the end of the century. The branches do not change during their recorded history, and if we understand evolution aright, they should not—for evolution is concentrated in rapid events of speciation, the production of new branches.

Homo sapiens is not the foreordained product of a ladder that was reaching toward our exalted estate from the start. We are merely the surviving branch of a once luxuriant bush.

7 The Child as Man's Real Father

PONCE DE LEON'S search for the fountain of youth continues in retirement villas of the sunshine state he discovered. Chinese alchemists once searched for the drug of deathlessness by allying the incorruptibility of flesh with the permanence of gold. How many of us would still make Faust's pact with the devil in exchange for perpetual life?

But our literature also records the potential problems of immortality. Wordsworth, in his famous ode, argued that childhood's bright vision of "splendor in the grass, of glory in the flower" can never be recaptured—though he counseled us to "grieve not, rather find strength in what remains behind." Aldous Huxley once devoted a novel—*After Many a Summer Dies the Swan*—to illustrating eternity's mixed blessing. With the consummate arrogance that only an American millionaire could display, Jo Stoyte sets out to purchase his immortality. Stoyte's hired scientist, Dr. Obispo, discovers that the fifth earl of Gonister has, by daily ingestion of carp guts, prolonged his life well beyond 200 years. They rush to England, break into the earl's guarded residence, and discover—to Stoyte's horror and Obispo's profound amusement—that the earl and his lover have grown up into apes. The horrid truth of our origin is out: We evolved by retaining the youthful features of our ancestors, a process known technically as neoteny (literally, "holding youth").

> A foetal ape that's had time to grow up," Dr. Obispo managed at last to say. "It's too good!" Laughter overtook him again. . . . Mr. Stoyte seized him by the shoulder and violently shook him. . . . "What's happened to them?" "Just time," said Dr. Obispo airily. . . . the foetal anthropoid was able to come to maturity. . . . without moving from where he was sitting, the Fifth Earl urinated on the floor.

Aldous Huxley got his theme from the "fetalization theory" proposed in the 1920s by the Dutch anatomist Louis Bolk (and probably transmitted to him by brother Julian, who had been doing some important research on delayed metamorphosis in amphibians). Bolk based his idea on the impressive list of features that we hold in common with the juvenile—but not the adult—stages of other primates or of mammals in general. The list includes, among more than twenty important characters, the following:

1. Our rounded, bulbous cranium—house of our larger brain. Embryonic apes and monkeys have a similar cranium, but the brain grows so much more slowly than the rest of the body (see essays 22 and 23) that the cranial vault becomes lower and relatively smaller in adults. Our brain itself probably achieved large size by the retention of rapid fetal growth rates.

2. Our "juvenile" face—straight profile, small jaws and teeth, weak brow ridges. The equally small jaws of juvenile apes grow relatively faster than the rest of the skull, forming a pronounced muzzle in adults.

3 . Position of the foramen magnum—the hole in our skull base from which the spinal cord emerges. As in the embryos of most mammals, our foramen magnum lies underneath our skull, pointing downward. Our skull is mounted on top of our backbone, and we look forward when standing upright. In other mammals, this embryonic location changes as the foramen moves to a position behind the skull pointing backward. This is suited for four-footed life since the head is now mounted in front of the vertebrae and the eyes are directed forward. The three morphological features most often cited as marks of humanity are our large brain, our small jaws and

our upright posture. The retention of juvenile features may have played an important role in evolving all of them.

4. Late closure of the skull sutures and other marks of delayed calcification of the skeleton. Babies have a large "soft spot," and the sutures between our skull bones do not fully close until well after adulthood. Thus, our brain can continue its pronounced postnatal expansion. (In most other mammals, the brain is nearly complete at birth and the skull is fully ossified.) A leading primate anatomist has remarked: "Though man grows *in utero* to larger sizes than any other primate, his skeletal maturation has progressed less at birth than in any monkey or ape for which relevant information has become available." Only in humans are the ends of long bones and digits still entirely cartilaginous at birth.

5. Ventral pointing of the vaginal canal in women. We copulate most comfortably face to face because we are built to do it that way. The vaginal canal also points forward in mammalian embryos, but it rotates back in adults, and males mount from the rear.

6. Our strong, unrotated, nonopposable big toe. The big toe of most primates begins as ours, in conjunction with its neighbors, but it rotates to the side and opposes the others for efficient grasping. By retaining a juvenile trait to yield a stronger foot for walking, our upright posture is enhanced.

Bolk's list was impressive (this is only a small part of it), but he tied it to a theory that doomed his observations to oblivion and gave Aldous Huxley his anti-Faustian metaphor. Bolk proposed that we evolved by an alteration in our hormonal balance that delayed development as a whole. He wrote:

> If I wished to express the basic principle of my ideas in a somewhat strongly worded sentence, I would say that man, in his bodily development, is a primate fetus that has become sexually mature.

Or, to quote Aldous Huxley again:

> There's a kind of glandular equilibrium. . . . Then a mutation comes along and knocks it sideways. You get a new equilibrium that happens to retard the developmental rate. You grow up; but you do it so slowly that

you're dead before you've stopped being like your great-great-grandfather's fetus.

Bolk did not shrink from the obvious implication. If we owe all our distinctive features to a hormonal brake on development, then that brake might be easily released: "You will note," writes Bolk, "that a number of what we might call pithecoid features dwell within us in latent condition, waiting only for the falling away of the retarding forces to become active again."

What a tenuous position for the crown of creation! An ape arrested in its development, holding the spark of divinity only through a chemical brake placed upon its glandular development.

Bolk's mechanism never won much support, but it began to grow in absurdity as modern Darwinian theory became established during the 1930s. How could a simple hormonal change produce such a complicated morphological response? Not all our features are retarded (long legs, for example), and those that are display varying degrees of delay. Organs evolve separately in response to differing adaptive requirements—a concept we call mosaic evolution. Unfortunately, Bolk's excellent observations were buried under the barrage of justified criticism for his fanciful mechanism. The theory of human neoteny is now usually relegated to a paragraph or two in anthropology textbooks. Yet I believe that it is fundamentally correct; an essential, if not dominant, theme in human evolution. But how can we rescue Bolk's observations from his theory?

If we must base our argument upon the list of neotenic features, then we are lost. The concept of mosaic evolution dictates that organs will evolve in different ways to meet varying selective pressures. Supporters of neoteny list their features, opponents tote theirs, and a stalemate is quickly reached. Who is to say which features are "more fundamental"? For example, one recent supporter of neoteny has written: "Most animals show retardation in some features, acceleration in others. . . . On balance, I think that in man, as compared with other primates, the slowing down far outweighs the speeding up." But a detractor proclaims: "the neotenic characters . . . are secondary consequences of the nonneotenic key characters." The validation of neoteny as fundamental requires more than an impressive list of retarded characters; it must be justified as an expected result of processes acting in human evolution.

The notion of neoteny achieved its initial fame as a way of opposing the theory of recapitulation, a dominant idea in late nineteenth century biology. The theory of recapitulation proclaimed that animals repeat the adult stages of their ancestors during their own embryonic and postnatal growth —ontogeny recapitulates phylogeny, in that mystical phrase we all learned in high school biology. (Recapitulationists argued that our embryonic gill slits represented the adult fish from which we descended.) If recapitulation were generally true—which it is not—then features would have to be *accelerated* during evolutionary history, for adult characters of ancestors can become the juvenile stages of descendants only if their development is speeded up. But neotenic characters are *retarded* since juvenile features of ancestors are delayed to appear in the adult stages of descendants. Thus, there is a general correspondence between accelerated development and recapitulation on the one hand and delayed development and neoteny on the other. If we can demonstrate a general delay of development in human evolution, then neoteny in key features becomes an expectation, not just an empirical tabulation.

I do not think that retardation can be denied as a basic event in human evolution. First of all, primates in general are retarded with respect to most other mammals. They live longer and mature more slowly than other mammals of comparable body size. The trend continues throughout the evolution of primates. Apes are generally larger, mature more slowly, and live longer than monkeys and prosimians. The course and tempo of our lives has slowed up even more dramatically. Our gestation period is only slightly longer than that of apes, but our babies are born much heavier—presumably because we retain our rapid fetal growth rates. I have already commented on the delay in ossification of our bones. Our teeth erupt later, we mature later, and we live longer. Many of our systems continue to grow long after comparable organs have ceased in other primates. At birth, the brain of a rhesus monkey is 65 percent of its final size, a chimpanzee's is 40.5 percent, but we attain only 23 percent. Chimps and gorillas reach 70 percent of final brain size early in their first year; we do not attain this value until early in our third year. W. M. Krogman, our leading expert in child growth, has written: "Man has absolutely the most protracted period of infancy, childhood and juvenility of all forms of life, i.e., he is a neotenous or long-growing animal. Nearly thirty percent of his entire life-span is devoted to growing."

This slowdown of our development does not guarantee that we will retain juvenile proportions as adults. But since neoteny and retarded development are generally linked, retardation does provide a mechanism for the easy retention of any juvenile feature that suits the adult life style of descendants. In fact, juvenile features are a storehouse of potential adaptations for descendants, and they can be utilized easily if development is strongly retarded in time (viz, the nonopposable big toe and small face of fetal primates—as discussed earlier). In our case, the "availability" of juvenile features clearly controlled the pathway to many of our distinctive adaptations.

But what is the adaptive significance of retarded development itself? The answer to this question probably lies in our social evolution. We are preeminently a learning animal. We are not particularly strong, swift, or well designed; we do not reproduce rapidly. Our advantage lies in our brain, with its remarkable capacity for learning by experience. To enhance our learning, we have lengthened our childhood by delaying sexual maturation with its adolescent yearning for independence. Our children are tied for longer periods to their parents, thus increasing their own time of learning and strengthening family ties as well.

This argument is an old one, but it wears well. John Locke (1689) praised our lengthy childhood for keeping parents together: "Wherein one cannot but admire the wisdom of the great Creator who . . . hath made it necessary that society of man and wife should be more lasting than that of male and female among other creatures, that so their industry might be encouraged, and their interest better united, to make provision and lay up goods for their common issue." But Alexander Pope (1735) said it even better, and in heroic couplets to boot:

> The beast and bird their common charge attend
> The mothers nurse it, and the sires defend
> The young dismissed, to wander earth and air,
> There stops the instinct, and there ends the care.
> A longer care man's helpless kind demands,
> That longer care contracts more lasting bands.

8 Human Babies as Embryos

MEL ALLEN, that irrepressible emcee of Yankee baseball during my youth,[3] finally aroused my displeasure by overenthusiastic endorsement of his sponsors. I never balked when he referred to home runs as "Ballantine blasts," but my patience was strained one afternoon when DiMaggio missed the left field foul pole by an inch and Allen exclaimed: "Foul by the ash on a White Owl cigar." I hope that I won't inspire any similar displeasure by confessing that I read and enjoy *Natural History* and that I even sometimes get an idea for an essay from its articles.

In the November 1975 issue, my friend Bob Martin wrote a piece on strategies of reproduction in primates. He focused upon the work of one of my favorite scientists—the idiosyncratic Swiss zoologist Adolf Portmann. In his voluminous studies, Portmann has identified two basic patterns in the reproductive strategies of mammals. Some mammals, usually designated by us as "primitive," have brief gestations and give birth to large litters of poorly developed young (tiny,

---

3 | I depart from my introductory promise to excise all topical references to the original source of these essays—my monthly column in *Natural History Magazine.* For where else will I ever have the opportunity to pay tribute to the man who ranks second only to my father for sheer volume of attention during my youth; he and the Yankees brought me so much pleasure (I even own a ball that DiMaggio fouled off one day).

hairless, helpless, and with unopened eyes and ears). Life-spans are short, brains small (relative to body size), and social behavior not well developed. Portmann refers to this pattern as altricial. On the other hand, many "advanced" mammals have long gestations, long life-spans, big brains, complex social behavior, and give birth to a few, well-developed babies capable, at least in part, of fending for themselves at birth. These traits mark the precocial mammals. In Portmann's vision of evolution as a process leading inexorably upward to greater spiritual development, the altricial pattern is primitive and preparatory to the higher precocial type that evolves along with enlarged brains. Most English-speaking evolutionists would reject this interpretation and link the basic patterns to immediate requirements of different modes of life. (I often exploit these essays to vent my own prejudice against equating evolution with "progress.") The altricial pattern, Martin argues, seems to correlate with marginal, fluctuating, and unstable environments in which animals do best by making as many offspring as they possibly can—so that some can weather the harshness and uncertainty of resources. The precocial pattern fits better with stable, tropical environments. Here, with more predictable resources, animals can invest their limited energy in a few, well-developed offspring.

Whatever the explanation, no one will deny that primates are the archetypical precocial mammals. Relative to body sizes, brains are biggest and gestation times and life-spans are longest among mammals. Litter size, in most cases, has been reduced to the absolute minimum of one. Babies are well developed and capable at birth. However, although Martin doesn't mention it, we encounter one obviously glaring and embarrassing exception—namely us. We share most of the precocial characters with our primate cousins—long life, large brains, and small litters. But our babies are as helpless and undeveloped at birth as those of most altricial mammals. In fact, Portmann himself refers to human babies as "secondarily altricial." Why did this most precocial of all species in some traits (notably the brain) evolve a baby far less developed and more helpless than that of its primate ancestors?

I will propose an answer to this question that is bound to strike most readers as patently absurd: Human babies are born as embryos, and embryos they remain for about the first nine months of life. If women gave birth when they "should" —after a gestation of about a year and a half—our babies would share the standard precocial features of other primates. This is Portmann's position, developed in a series of German articles during the 1940s and essentially unknown in this country. Ashley Montagu reached the same conclusion independently in a paper published in the *Journal of the American Medical Association* in October 1961. Oxford psychologist R. E. Passingham championed it in a piece published late in 1975 in the technical journal *Brain, Behavior and Evolution.* I also cast my lot with this select group in regarding the argument as basically correct.

The initial impression that such an argument can only be arrant nonsense arises from the length of human gestation. Gorillas and chimps may not be far behind, but human gestation is still the longest among primates. How then can I claim that human neonates are embryos because they are born (in some sense) too soon? The answer is that planetary days may not provide an appropriate measure of time in all biological calculations. Some questions can only be treated properly when time is measured relatively in terms of an animal's own metabolism or developmental rate. We know, for example, that mammalian life-spans vary from a few weeks to more than a century. But is this a "real" distinction in terms of a mammal's own perception of time and rate? Does a rat really live "less" than an elephant? Laws of scaling dictate that small, warm-blooded animals live at a faster pace than larger relatives (see essays 21 and 22). The heart beats more rapidly and metabolism proceeds at a greatly elevated rate. In fact, for several criteria of relative time, all mammals live about the same amount. All, for example, breathe about the same number of times during their lives (small, short-lived mammals breathe more rapidly than larger, slow metabolizers).

In astronomical days, human gestation is long, but relative to human developmental rates, it is truncated and abbreviated. In the previous essay, I argued that a (if not the) major feature of human evolution has been the marked slowing up of our development. Our brains grow more slowly and for a longer time than those of other primates, our bones ossify much later, and the period of our childhood is greatly extended. In fact, we never reach the levels of development attained by most primates. Human adults retain, in several important respects, the juvenile traits of ancestral primates—an evolutionary phenomenon called neoteny.

Compared with other primates, we grow and develop at a snail's pace; yet our gestation period is but a few days longer than that of gorillas and chimpanzees. Relative to our own developmental rate, our gestation has been markedly shortened. If length of gestation had slowed down as much as the rest of our growth and development, human babies would be born anywhere from seven to eight months (Passingham's estimate) to a year (Portmann and Ashley Montagu's estimate) after the nine months actually spent *in utero.*

But am I not indulging in mere metaphor or trick of phrase in designating the human baby as "still an embryo"? I have just raised two of my own past this tender age, and have experienced all the joy and mystery of their mental and physical development—things that could never happen in a dark, confining womb. Still, I side with Portmann when I consider the data on their physical growth, for during their first year, human babies share the growth patterns of primate and mammalian fetuses, not of other primate babies. (The identification of certain growth patterns as either fetal or postnatal is not arbitrary. Postnatal development is not a mere prolongation of fetal tendencies; birth is a time of marked discontinuity in many features.) Human neonates, for example, have not yet ossified the ends of limb bones or fingers; ossification centers are usually entirely absent in the finger bones of newborn humans. This level of ossification corresponds to the eighteenth fetal week of macaque monkeys. When macaques are born at twenty-four weeks, their limb bones are ossified to an extent not reached by humans until years after birth. More crucially, our brains continue to grow at rapid, fetal rates after birth. The brains of many mammals are essentially fully formed at birth. Other primates extend brain development into early postnatal growth. The brain of a human baby is only one-fourth its final size at birth. Passingham writes: "Man's brain does not reach the proportion found for the chimpanzee at birth until around six months after birth. This time corresponds quite well with the time at which man would be expected to be born if his gestation period were as high a proportion of his development and life-span as it is in apes."

A. H. Schultz, one of the greatest primate anatomists of the century, summarized his comparative study of growth in primates by stating: "It is evident that human ontogeny is not unique in regard to the duration of life in utero, but that it has become highly specialized in the striking postponement of the completion of growth and of the onset of senility."

But why are human babies born before their time? Why has evolution extended our general development so greatly, but held our gestation time in check, thereby giving us an essentially embryonic baby? Why was gestation not equally prolonged with the rest of development? In Portmann's spiritual view of evolution, this precocious birth must be a function of mental requirements. He argues that humans, as learning animals, need to leave the dark, unchallenging womb to gain access, as flexible embryos, to the rich extrauterine environment of sights, smells, sounds, and touches.

But I believe (along with Ashley Montagu and Passingham) that a more important reason lies in a consideration that Portmann dismisses contemptuously as coarsely mechanical and materialistic. From what I have seen (although I cannot know for sure), human birth is a joyful experience when properly rescued from arrogant male physicians who seem to want total control over a process they cannot experience. Nonetheless, I do not think it can be denied that human birth is difficult compared with that of most other mammals. To put it rather grossly, it's a tight squeeze. We know that female primates can die in attempted childbirth when fetal heads are too large to pass through the pelvic canal. A. H. Schultz illustrates the stillborn fetus of a hamadryas baboon and the pelvic canal of its dead mother; the embryo's head is a good deal larger than the canal. Schultz concludes that fetal size is near its limit in this species: "While selection undoubtedly tends to favor large diameters of the female pelvis, it must also act against any prolongation of gestation or at least against unduly large newborns."

There are not, I am confident, many human females who could give birth successfully to a year-old baby.

The culprit in this tale is our most important evolutionary specialization, our large brain. In most mammals, brain growth is entirely a fetal phenomenon. But since the brain never gets very large, this poses no problem for birth. In larger-brained monkeys, growth is delayed somewhat to permit postnatal enlargement of the brain, but relative times of gestation need not be altered. Human brains, however, are so large that another strategy must be added for successful birth—gestation must be shortened relative to general development, and birth must occur when the brain is only one-fourth its final size.

Our brain has probably reached the end of its increase in size. The paramount trait of our evolution has finally limited its own potential for future growth. Barring some radical redesign of the female pelvis, we will have to make do with the brains we have if we want to be born at all. But, no matter. We can happily spend the next several millennia learning what to do with an immense potential that we have scarcely begun to understand or exploit.

3 Odd Organisms and Evolutionary Exemplars

9 The Misnamed, Mistreated, and Misunderstood Irish Elk

Nature herself seems by the vast magnitude and stately horns, she has given this creature, to have singled it out as it were, and showed it such regard, with a design to distinguish it remarkably from the common herd of all other smaller quadrupeds.

THOMAS MOLYNEUX, *1697*

THE IRISH ELK, the Holy Roman Empire, and the English horn form a strange ensemble indeed. But they share the common distinction of their completely inappropriate names. The Holy Roman Empire, Voltaire tells us, was neither holy, nor Roman, nor an empire. The English horn is a continental oboe; the original versions were curved, hence "angular" (corrupted to English) horn. The Irish Elk was neither exclusively Irish, nor an elk. It was the largest deer that ever lived. Its enormous antlers were even more impressive. Dr. Molyneux marveled at "these spacious horns" in the first published description of 1697. In 1842, Rathke described them in a language unexcelled for the expression of enormity as *bewunderungswuerdig.* Although the Guiness book of world records ignores fossils and honors the American moose, the antlers of the Irish Elk have never been exceeded, or even approached, in the history of life. Reliable estimates of their total span range up to 12 feet. This figure seems all the more impressive when we recognize that the antlers were probably shed and regrown annually, as in all other true deer.

Fossil antlers of the giant deer have long been known in Ireland, where they occur in lake sediments underneath peat deposits. Before attracting the attention of scientists, they had been used as gateposts, and even as a temporary bridge to span a rivulet in County Tyrone. One story, probably apocryphal, tells of a huge bonfire made of their bones and antlers in County Antrim to celebrate the victory over Napoleon at Waterloo. They were called elk because the European moose (an "elk" to Englishmen) was the only familiar animal with antlers that even approached those of the giant deer in size.

The first known drawing of giant deer antlers dates from 1588. Nearly a century later, Charles II received a pair of antlers and (according to Dr. Molyneux) "valued them so highly for their prodigious largeness" that he set them up in the horn gallery of Hampton Court, where they "so vastly exceed" all others in size "that the rest appear to lose much of their curiosity."

Ireland's exclusive claim vanished in 1746 (although the name stuck) when a skull and antlers were unearthed in Yorkshire, England. The first continental discovery followed in

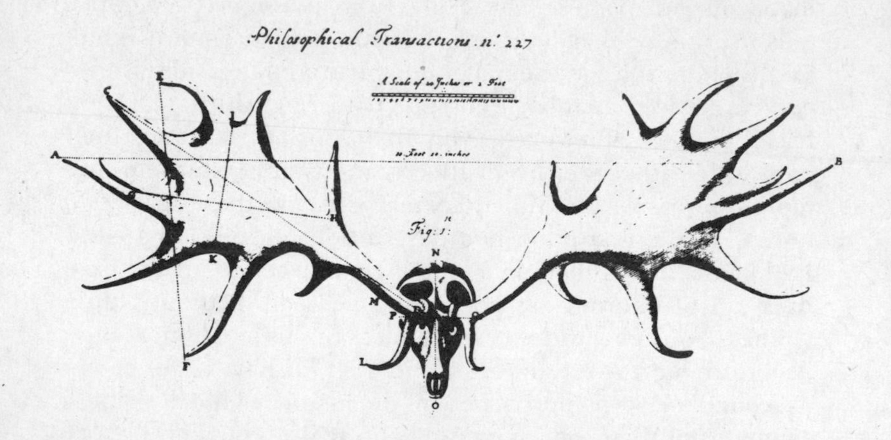

A drawing of the giant deer in Thomas Molyneux's 1697 article shows the antlers incorrectly rotated forward ninety degrees.

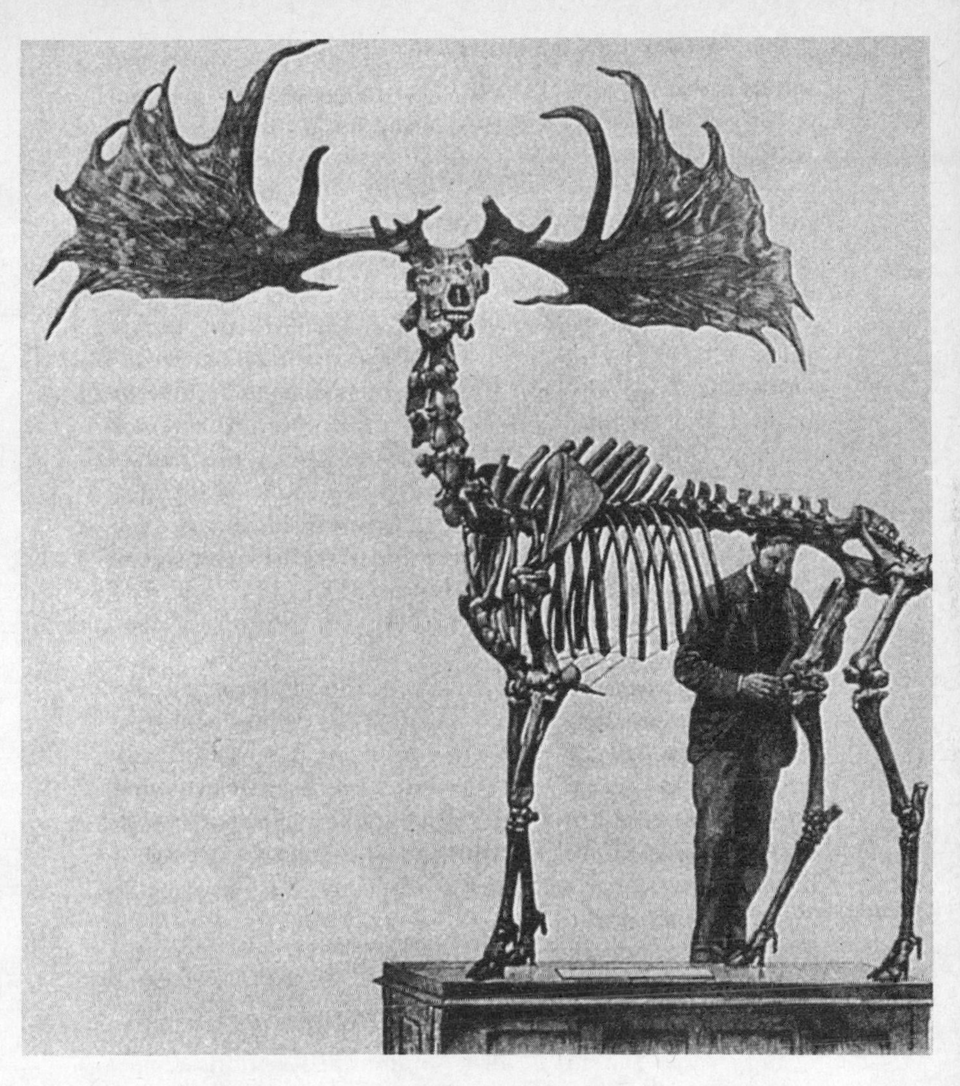

A worthy predecessor of the author measures the other end of an Irish Elk. Figure originally published by J. G. Millais in 1897.

1781 from Germany, while the first complete skeleton (still standing in the museum of Edinburgh University) was exhumed from the Isle of Man in the 1820s.

We now know that the giant deer ranged as far east as Siberia and China and as far south as northern Africa. Specimens from England and Eurasia are almost always fragmentary, and nearly all the fine specimens that adorn so many museums throughout the world come from Ireland. The giant deer evolved during the glacial period of the last few million years and may have survived to historic times in continental Europe, but it became extinct in Ireland about 11,000 years ago.

"Among the fossils of the British empire," wrote James Parkinson in 1811, "none are more calculated to excite astonishment." And so it has been throughout the history of paleontology. Putting aside both the curious anecdotes and the sheer wonder that immensity always inspires, the importance of the giant deer lies in its contribution to debates about evolutionary theory. Every great evolutionist has used the giant deer to defend his favored views. The controversy has centered around two main issues: (1) Could antlers of such bulk be of any use? and (2) Why did the giant deer become extinct?

Since debate on the Irish Elk has long centered on the reasons for its extinction, it is ironic that the primary purpose of Molyneux's original article was to argue that it must still be alive. Many seventeenth-century scientists maintained that the extinction of any species would be inconsistent with God's goodness and perfection. Dr. Molyneux's article of 1697 begins:

> That no real species of living creatures is so utterly extinct, as to be lost entirely out of the World, since it was first created, is the opinion of many naturalists; and 'tis grounded on so good a principle of Providence taking care in general of all its animal productions, that it deserves our assent.

Yet the giant deer no longer inhabited Ireland, and Molyneux was forced to search elsewhere. After reading traveler's reports of antler size in the American moose, he concluded that the Irish Elk must be the same animal; the tendency toward exaggeration in such accounts is apparently universal and timeless. Since he could find neither figure nor an accurate description of the moose, his conclusions are not as absurd as modern knowledge would indicate. Molyneux attributed the giant deer's demise in Ireland to an "epidemick distemper," caused by "a certain ill constitution of air."

For the next century arguments raged along Molyneux's line—to which modern species did the giant deer belong? Opinion was equally divided between the moose and the reindeer.

As eighteenth-century geologists unraveled the fossil record of ancient life, it became more and more difficult to argue that the odd and unknown creatures revealed by fossils were all still living in some remote portion of the globe. Perhaps God had not created just once and for all time; perhaps He had experimented continually in both creation and destruction. If so, the world was surely older than the six thousand years that literalists allowed.

The question of extinction was the first great battleground of modern paleontology. In America, Thomas Jefferson maintained the old view, while Georges Cuvier, the great French paleontologist, was using the Irish Elk to prove that extinction did occur. By 1812 Cuvier had resolved two pressing issues: by minute anatomical description, he proved that the Irish Elk was not like any modern animal; and by placing it among many fossil mammals with no modern counterparts, he established the fact of extinction and set the basis for a geologic time scale.

Once the fact of extinction had been settled, debate moved to the time of the event: in particular, had the Irish elk survived the flood? This was no idle matter, for if the flood or some previous catastrophe had wiped out the giant deer, then its demise had natural (or supernatural) causes. Archdeacon Maunsell, a dedicated amateur, wrote in 1825: "I apprehended they must have been destroyed by some overwhelming deluge." A certain Dr. MacCulloch even believed that the fossils were found standing erect, noses elevated—a final gesture to the rising flood, as well as a final plea: don't make waves.

If, however, they had survived the flood, then their exterminating angel could only have been the naked ape himself. Gideon Mantell, writing in 1851, blamed Celtic tribes; in

1830, Hibbert implicated the Romans and the extravagant slaughters of their public games. Lest we assume that our destructive potential was recognized only recently, Hibbert wrote in 1830: "Sir Thomas Molyneux conceived that a sort of distemper, or pestilential murrain, might have cut off the Irish Elks. . . . It is, however, questionable, if the human race has not occasionally proved as formidable as a pestilence in exterminating from various districts, whole races of wild animals."

In 1846, Britain's greatest paleontologist, Sir Richard Owen, reviewed the evidence and concluded that in Ireland at least, the giant deer had perished before man's arrival. By this time, Noah's flood as a serious geologic proposition had passed from the scene. What then had wiped out the giant deer?

Charles Darwin published the *Origin of Species* in 1859. Within ten years virtually all scientists had accepted the *fact* of evolution. But the debate about causes and mechanisms was not resolved (in Darwin's favor) until the 1940s. Darwin's theory of natural selection requires that evolutionary changes be adaptive—that is, that they be useful to the organism. Therefore, anti-Darwinians searched the fossil record for cases of evolution that could not have benefited the animals involved.

The theory of orthogenesis became a touchstone for anti-Darwinian paleontologists, for it claimed that evolution proceeded in straght lines that natural selection could not regulate. Certain trends, once started, could not be stopped even if they led to extinction. Thus certain oysters, it was said, coiled their valves upon each other until they sealed the animal permanently within; saber–toothed "tigers" could not stop growing their teeth or mammoths their tusks.

But by far the most famous example of orthogenesis was the Irish Elk itself. The giant deer had evolved from small forms with even smaller antlers. Although the antlers were useful at first, their growth could not be contained and, like the sorceror's apprentice, the giant deer discovered only too late that even good things have their limits. Bowed by the weight of their cranial excrescences, caught in the trees or mired in the ponds, they died. What wiped out the Irish Elk? They themselves or, rather, their own antlers did.

In 1925, the American paleontologist R. S. Lull invoked the giant deer to attack Darwinism: "Natural selection will not account for overspecialization, for it is manifest that, while an organ can be brought to the point of perfection by selection, it would never be carried to a condition where it is an actual menace to survival . . . [as in] the great branching antlers of the extinct Irish deer."

Darwinians, led by Julian Huxley, launched a counterattack in the 1930s. Huxley noted that as deer get larger—either during their own growth or in the comparison of related adults of different sizes—the antlers do not increase in the same proportion as body size; they increase faster, so that the antlers of large deer are not only absolutely larger but also relatively larger than those of small deer. For such regular and orderly change of shape with increasing size, Huxley used the term allometry.

Allometry provided a comfortable explanation for the giant deer's antlers. Since the Irish Elk had the largest body size of any deer, its relatively enormous antlers could have been a simple result of the allometric relationship present among all deer. We need only assume that increased body size was favored by natural selection; the large antlers might have been an automatic consequence. They might even have been slightly harmful in themselves, but this disadvantage was more than compensated by the benefits of larger size, and the trend continued. Of course, when problems of larger antlers outweighed the advantages of larger bodies, the trend would cease since it could no longer be favored by natural selection.

Almost every modern textbook of evolution presents the Irish Elk in this light, citing the allometric explanation to counter orthogenetic theories. As a trusting student, I had assumed that such constant repetition must be firmly based on copious data. Later I discovered that textbook dogma is self-perpetuating; therefore, three years ago I was disappointed, but not really surprised, to discover that this widely touted explanation was based on no data whatsoever. Aside from a few desultory attempts to find the largest set of antlers, no one had ever measured an Irish Elk. Yardstick in hand, I resolved to rectify this situation.

The National Museum of Ireland in Dublin has seventeen specimens on display and many more, piled antler upon antler, in a nearby warehouse. Most large museums in western Europe and America own an Irish Elk, and the giant deer adorns many trophy rooms of English and Irish gentry. The largest antlers grace the entranceway to Adare Manor, home of the Earl of Dunraven. The sorriest skeleton sits in the cellar of Bunratty Castle, where many merry and slightly inebriated tourists repair for coffee each evening after a medieval banquet. This poor fellow, when I met him early the morning after, was smoking a cigar, missing two teeth, and carrying three coffee cups on the tines of his antlers. For those who enjoy invidious comparisons, the largest antlers in America are at Yale; the smallest in the world at Harvard.

To determine if the giant deer's antlers increased allometrically, I compared antler and body size. For antler size, I used a compounded measure of antler length, antler width, and the lengths of major tines. Body length, or the length and width of major bones, might be the most appropriate measure of body size, but I could not use it because the vast majority of specimens consist only of a skull and its attached antlers. Moreover, the few complete skeletons are invariably made up of several animals, much plaster, and an occasional ersatz (the first skeleton in Edinburgh once sported a horse's pelvis). Skull length therefore served as my measure of overall size. The skull reaches its final length at a very early age (all my specimens are older) and does not vary thereafter; it is, therefore, a good indicator of body size. My sample included seventy-nine skulls and antlers from museums and homes in Ireland, Britain, continental Europe, and the United States.

My measurements showed a strong positive correlation between antler size and body size, with the antlers increasing in size two and one-half times faster than body size from small to large males. This is not a plot of individual growth; it is a relationship among adults of different body size. Thus, the allometric hypothesis is affirmed. If natural selection favored large deer, then relatively larger antlers would appear as a correlated result of no necessary significance in itself.

Yet, even as I affirmed the allometric relationship, I began to doubt the traditional explanation—for it contained a curious remnant of the older, orthogenetic view. It assumed that

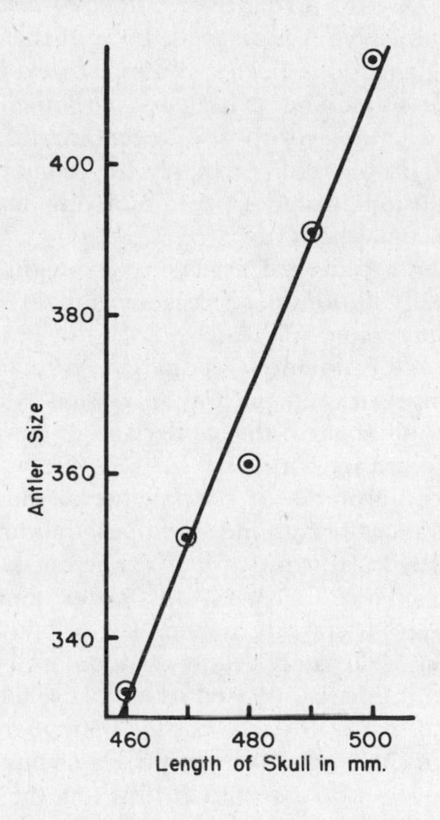

Graph showing relative increase in antler size with increasing skull length in Irish Elks. Each point is the average for all skulls in a 10 mm. interval of length; the actual data include 81 individuals. Antler size increases more than 2½ times as fast as skull length—a line with a slope of 1.0 (45 degree angle with the x-axis) would indicate equal rates of increase on these logarithmic scales. The slope here is obviously very much higher.

the antlers are not adaptive in themselves and were tolerated only because the advantages of increased body size were so great. But why must we assume that the immense antlers had no primary function? The opposite interpretation is equally possible: that selection operated primarily to increase antler size, thus yielding increased body size as a secondary consequence. The case for inadaptive antlers has never rested on more than subjective wonderment born of their immensity.

Views long abandoned often continue to exert their influence in subtle ways. The orthogenetic argument lived on in the allometric context proposed to replace it. I believe that the supposed problem of "unwieldy" or "cumbersome" antlers is an illusion rooted in a notion now abandoned by students of animal behavior.

To nineteenth-century Darwinians, the natural world was a cruel place. Evolutionary success was measured in terms of battles won and enemies destroyed. In this context, antlers were viewed as formidable weapons to be used against predators and rival males. In his *Descent of Man* (1871), Darwin toyed with another idea: that antlers might have evolved as ornaments to attract females. "If, then, the horns, like the splendid accouterments of the knights of old, add to the noble appearance of stags and antelopes, they may have been modified partly for this purpose." Yet he quickly added that he had "no evidence in favor of this belief," and went on to interpret antlers according to the "law of battle" and their advantages in "reiterated deadly contests." All early writers assumed that the Irish Elk used its antlers to kill wolves and drive off rival males in fierce battle. To my knowledge this view has been challenged only by the Russian paleontologist L. S. Davitashvili, who asserted in 1961 that the antlers functioned primarily as courtship signals to females.

Now, if antlers are weapons, the orthogenetic argument is appealing, for I must admit that ninety pounds of broad-palmed antler, regrown annually and spanning twelve feet from tip to tip, seems even more inflated than our current military budget. Therefore, to preserve a Darwinian explanation, we must invoke the allometric hypothesis in its original form.

But what if antlers do not function primarily as weapons? Modern studies of animal behavior have generated an exciting concept of great importance to evolutionary biology: many structures previously judged as actual weapons or devices for display to females are actually used for ritualized combat among males. Their function is to prevent actual battle (with consequent injuries and loss of life) by establishing hierarchies of dominance that males can easily recognize and obey.

Antlers and horns are a primary example of structures used for ritualized behavior. They serve, according to Valerius Geist, as "visual dominance-rank symbols." Large antlers confer high status and access to females. Since there can be no evolutionary advantage more potent than a guarantee of successful reproduction, selective pressures for larger antlers must often be intense. As more and more horned animals are observed in their natural environment, older ideas of deadly battle are yielding to evidence of purely ritualized display without body contact, or fighting in ways clearly designed to prevent bodily injury. This has been observed in red deer by Beninde and Darling, caribou by Kelsall, and in mountain sheep by Geist.

As devices for display among males, the enormous antlers of the Irish Elk finally make sense as structures adaptive in themselves. Moreover, as R. Coope of Birmingham University pointed out to me, the detailed morphology of the antlers can be explained, for the first time, in this context. Deer with broad-palmed antlers tend to show the full width of their antlers in display. The modern fallow deer (considered by many as the Irish Elk's nearest living relative) must rotate its head from side to side in order to show its palm. This would have created great problems for giant deer, since the torque produced by swinging ninety-pound antlers would have been immense. But the antlers of the Irish Elk were arranged to display the palm fully when the animal looked straight ahead. Both the unusual configuration and the enormous size of the antlers can be explained by postulating that they were used for display rather than for combat.

If the antlers were adaptive, why did the Irish Elk become

extinct (at least in Ireland)? The probable answer to this old dilemma is, I am afraid, rather commonplace. The giant deer flourished in Ireland for only the briefest of times—during the so-called Alleröd interstadial phase at the end of the last glaciation. This period, a minor warm phase between two colder epochs, lasted for about 1,000 years, from 12,000 to 11,000 years before the present. (The Irish Elk had migrated to Ireland during the previous glacial phase when lower sea levels established a connection between Ireland and continental Europe.) Although it was well adapted to the grassy, sparsely wooded, open country of Alleröd times, it apparently could not adapt either to the subarctic tundra that followed in the next cold epoch or to the heavy forestation that developed after the final retreat of the ice sheet.

Extinction is the fate of most species, usually because they fail to adapt rapidly enough to changing conditions of climate or competition. Darwinian evolution decrees that no animal shall actively develop a harmful structure, but it offers no guarantee that useful structures will continue to be adaptive in changed circumstances. The Irish Elk was probably a victim of its own previous success. *Sic transit gloria mundi.*

10 Organic Wisdom, or Why Should a Fly Eat Its Mother from Inside

SINCE MAN CREATED God in his own image, the doctrine of special creation has never failed to explain those adaptations that we understand intuitively. How can we doubt that animals are exquisitely designed for their appointed roles when we watch a lioness hunt, a horse run, or a hippo wallow? The theory of natural selection would never have replaced the doctrine of divine creation if evident, admirable design pervaded all organisms. Charles Darwin understood this, and he focused on features that would be out of place in a world constructed by perfect wisdom. Why, for example, should a sensible designer create only on Australia a suite of marsupials to fill the same roles that placental mammals occupy on all other continents? Darwin even wrote an entire book on orchids to argue that the structures evolved to insure fertilization by insects are jerry-built of available parts used by ancestors for other purposes. Orchids are Rube Goldberg machines; a perfect engineer would certainly have come up with something better.

This principle remains true today. The best illustrations of adaptation by evolution are the ones that strike our intuition as peculiar or bizarre. Science is not "organized common sense"; at its most exciting, it reformulates our view of the world by imposing powerful theories against the ancient, anthropocentric prejudices that we call intuition.

Consider, for example, the cecidomyian gall midges. These tiny flies conduct their lives in a way that tends to evoke feelings of pain or disgust when we empathize with them by applying the inappropriate standards of our own social codes.

Cecidomyian gall midges can grow and develop along one of two pathways. In some situations, they hatch from eggs, go through a normal sequence of larval and pupal molts, and emerge as ordinary, sexually reproducing flies. But in other circumstances, females reproduce by parthenogenesis, bringing forth their young without any fertilization by males. Parthenogenesis is common enough among animals, but the cecidomyians give it an interesting twist. First of all, the parthenogenetic females stop at an early stage of development. They never become normal, adult flies, but reproduce while they are still larvae or pupae. Secondly, these females do not lay eggs. The offspring develop live within their mother's body—not supplied with nutrient and packaged away in a protected uterus but right within the mother's tissues, eventually filling her entire body. In order to grow, the offspring devour their mother from the inside. A few days later, they emerge, leaving a chitinous shell as the only remains of their only parent. And within two days, their own developing children are beginning, literally, to eat them up.

Micromalthus debilis, an unrelated beetle, has evolved an almost identical system with a macabre variation. Some parthenogenetic females give birth to a single male offspring. This larva attaches to his mother's cuticle for about four or five days, then inserts his head into her genital aperture and devours her. Greater love hath no woman.

Why has such a peculiar mode of reproduction evolved? For it is unusual even among insects, and not only by the irrelevant standards of our own perceptions. What is the adaptive significance of a mode of life that so strongly violates our intuitions about good design?

To answer these questions, we proceed by the usual mode of argument in evolutionary studies: the comparative method. (Louis Agassiz did not act capriciously when he gave to the building in which I work the name that has puzzled so many generations of visitors to Harvard—the Museum of Comparative Zoology.) We must find an object for comparison that is genetically similar, but adapted to a different mode of life. Fortunately, the complex life cycle of cecidomyians provides us with a key. We do not have to compare the asexual, larval mother with a related species of uncertain affinity and genetic resemblance; we may contrast it with the genetically identical, alternate form of the same species—the normal, sexual fly. What then is different about the ecology of parthenogenetic and normal forms?

The cecidomyians feed and dwell on fungi, usually mushrooms. The mobile, normal fly fills the role of discoverer: it finds the new mushroom. Its offspring, now living on a superabundant food resource, reproduce asexually as larvae or pupae and become the flightless, feeding form of the species (a mushroom can support hundreds of these tiny flies). We know that parthenogenetic reproduction will continue as long as food is abundant. One investigator produced 250 consecutive larval generations by supplying enough food and preventing crowding. In nature, however, the mushroom is eventually used up.

H. Ulrich and his coworkers have studied the sequence of changes in response to decreasing food in the species *Mycophila speyeri.* When they have abundant food, parthenogenetic mothers generate all female broods in four to five days. As the supply of food diminishes, all male and mixed male and female broods develop. If female larvae are not fed at all, they grow into normal flies.

These correlations have a fairly unambiguous adaptive basis. The flightless, parthenogenetic female stays on the mushroom and feeds. When it exhausts its resource, it produces winged descendants to find new mushrooms. But this only scratches the surface of our dilemma, for it does not address our central question: Why reproduce so quickly as a larva or pupa, and why self-destruct by a supreme sacrifice to one's children?

I believe that the solution to this dilemma lies in the phrase "so quickly." Traditional evolutionary theory concentrated on morphology in framing adaptive explanations. What, in this case, is the advantage to mushroom feeders of a persistent juvenile morphology in reproducing females? Traditional theory never found an answer because it had posed the wrong question. During the last fifteen years, the rise of theoretical population ecology has transformed the study of adaptation. Evolutionists have learned that organisms adapt not only by altering their size and shape but also by adjusting the timing of their lives and the energy invested in different activities (feeding, growth, and reproduction, for example). These adjustments are called "life history strategies."

Organisms evolve different life history strategies to fit differnt types of environments. Among theories that correlate strategy with environment, the theory of *r*- and *K*- selection, developed by R. H. MacArthur and E. O. Wilson in the mid-1960s, has surely been the most successful.

Evolution, as usually depicted in textbooks and reported in the popular press, is a process of inexorable improvement in form: animals are delicately "fine tuned" to their environment through constant selection of better-adapted shapes. But several kinds of environments do not call forth such an evolutionary response. Suppose that a species lives in an environment that imposes irregular, catastrophic mortality upon it (ponds that dry up, for example, or shallow seas ripped up by severe storms). Or suppose that food sources are ephemeral and hard to find, but superabundant once located. Organisms cannot fine tune themselves to such environments for there is nothing sufficiently stable to adjust to. Better in such a situation to invest as much energy as possible into reproduction—make as many offspring as you can, as quickly as possible, so that some will survive the catastrophe. Reproduce like hell while you have the ephemeral resource, for it will not last long and some of your progeny must survive to find the next one.

We refer to evolutionary pressures for the maximization of reproductive effort at the expense of delicate morphological adjustment as *r*-selection; organisms so adapted are *r*-strategists (*r* is the traditional measure of "intrinsic rate of increase in population size" in a set of basic, ecological equations). Species that live in stable environments, near the maximum population size that the environment can support, will gain nothing by producing hordes of poorly adjusted progeny.

Better to raise a few, finely tuned offspring. Such species are *K*-strategists (*K* is the measure of environmental "carrying capacity" in the same set of equations).

The parthenogenetic larval gall midges live in a classical *r*-environment. Mushrooms are few and far between, but superabundant when found by such a tiny fly. Cecidomyian gall midges therefore gain a selective advantage if they use newly discovered mushrooms for building up their population as rapidly as possible. What, then, is the most efficient way to build a population quickly? Should the midges simply lay more eggs or should they reproduce as early as possible during their lives? This general issue has inspired a large literature among mathematically inclined ecologists. In most situations, the key to rapid increase is *early* reproduction. A 10 percent decrease in age at first reproduction can often yield the same effect as a 100 percent increase in fecundity.

Finally, we can understand the peculiar reproductive biology of cecidomyian gall midges: they have simply evolved some remarkable adaptations for early reproduction and extremely short generation times. In so doing, they have become consummate *r*-strategists in their classical *r*-environment of ephemeral, superabundant resources. Thus, they reproduce while still larvae, and almost immediately after hatching, they begin to grow the next generation within themselves. In *Mycophila speyeri,* for example, the parthenogenetic *r*-strategist undergoes only one molt, reproduces as a true larva, and manufactures up to 38 offspring in five days. The normal, sexual adults require two weeks to develop. The larval reproducers maintain a phenomenal capacity for increase in population size. Within five weeks after its introduction into a commercial mushroom bed, *Mycophila speyeri* can reach a density of 20,000 reproductive larvae per square foot.

We may again pursue the comparative method to convince ourselves that this explanation makes sense. The cecidomyian pattern has been followed by other insects that inhabit a similar set of environments. Aphids, for example, feed on the sap of leaves. A leaf, to these tiny insects, is much like a mushroom to a gall midge—a large, ephemeral resource to be converted quickly into as many aphids as possible. Most aphids have alternate parthenogenetic forms—wingless and winged (they also have an overwintering, sexual form, which need not concern us here). As you have probably already guessed, the wingless form is a flightless feeder. Although it is not a larva, it retains many features of juvenile morphology. It also maintains a remarkable capacity for early reproduction. Embryonic development actually begins in a mother's body before her own birth, and two subsequent generations may be telescoped within each "grandmother." (Aphids, however, are not consumed by their offspring.) Their capacity for rapid increase in population size is legendary. If all its offspring lived to reproduce, a single female of *Aphis fabae* could produce 524 billion progeny in a year. Winged aphids develop more slowly when the leaf is used up. They fly off to a new leaf, where their offspring revert to the wingless form and begin their rapid cycling of generations.

What at first seemed so peculiar now seems eminently reasonable. It may even be an optimal strategy for certain environments. This much we cannot claim, for so many aspects of cecidomyian biology are entirely unknown. But we can point to the uncanny convergence upon the same strategy by a completely unrelated organism, the beetle *Micromalthus debilis.* This beetle lives and feeds in wet, rotting wood. When the wood dries out, the beetle develops a sexual form to search for new resources. The wood-dwelling, feeding form has evolved a set of adaptations that repeats the features of cecidomyians down to the most complex and peculiar detail. It also is parthenogenetic. It also reproduces at a morphologically juvenile stage. The young also develop within the mother's body and eventually devour her. Mothers also produce three types of broods: females only when food is abundant and males only or males and females when resources diminish.

We humans with our slow development (see essay 7), extended gestation, and minimal litter size are consummate *K*-strategists and we may look askance at the strategies of other organisms, but in their *r*-selective world the cecidomyians are surely doing something right.

11 Of Bamboos, Cicadas, and the Economy of Adam Smith

NATURE USUALLY manages to outdo even the most fanciful of human legends. Sleeping Beauty waited a hundred years for her prince. Bettelheim argues that her pricked finger represents the first bleeding of menstruation, her long sleep the lethargy of adolescence awaiting the onset of full maturity. Since the original Sleeping Beauty was inseminated by a king, rather than merely kissed by a prince, we may interpret her awakening as the beginning of sexual fulfillment (see B. Bettelheim, *The Uses of Enchantment,* A. Knopf, 1976, pp. 225–36).

A bamboo bearing the formidable name *Phyllostachys bambusoides* flowered in China during the year 999. Since then, with unerring regularity, it has continued to flower and set seed roughly every 120 years. *P. bambusoides* follows this cycle wherever it lives. In the late 1960s, Japanese stocks (themselves transplanted from China centuries before) set seed simultaneously in Japan, England, Alabama, and Russia. The analogy to Sleeping Beauty is not farfetched, for sexual reproduction follows more than a century of celibacy in these bamboos. But *P. bambusoides* departs from the Brothers Grimm in two important ways. The plants are not inactive during their 120 year vigil—for they are grasses, and they propagate asexually by producing new shoots from underground rhizomes. Also, they do not live happily ever after, for they die after setting seed—a long wait for a short end.

Ecologist Daniel H. Janzen of the University of Pennsylvania recounts the curious tale of *Phyllostachys* in a recent article, "Why bamboos wait so long to flower" *(Annual Review of Ecology and Systematics,* 1976). Most species of bamboo have shorter periods of vegetative growth between flowerings, but synchroneity of seeding is the rule, and very few species wait fewer than 15 years before flowering (some may wait for more than 150 years, but historical records are too sparse to permit firm conclusions).

The flowering of any species must be set by an internal, genetic clock, not imposed from without by some environmental clue. The unerring regularity of repetition supplies our best evidence for this assertion, for we do not know any environmental factor that cycles so predictably to yield the variety of clocks followed by more than a hundred species. Secondly, as mentioned above, plants of the same species flower simultaneously, even when transplanted half a world away from their native habitat. Finally, plants of the same species flower together, even if they have grown in very different environments. Janzen recounts the tale of a Burmese bamboo only half a foot high that had been burned down repeatedly by jungle fires, but flowered at the same time as its unhurt companions standing 40 feet tall.

How can a bamboo count the passing years? Janzen argues that it cannot be measuring stored food reserves because starved dwarfs flower at the same time as healthy giants. He speculates that the calendar "must be the annual or daily accumulation or degradation of a temperature-insensitive photosensitive chemical." He finds no basis for guessing whether the cycles of light are diurnal (day-night) or yearly (seasonal). As circumstantial evidence for implicating light as a clock, Janzen points out that no accurately cycling bamboo grows within 5 degrees of latitude from the equator—for variations in both days and seasons are minimized within this zone.

The flowering of bamboo recalls a tale of striking periodicity better known to most of us—the periodical cicada, or 17-year "locust." (Cicadas are not locusts at all, but large-bodied members of the order Homoptera, a group of predominantly small insects including aphids and their relatives; locusts, along with crickets and grasshoppers, form the order Orthoptera.) The story of periodical cicadas is even more amazing than most people realize: for 17 years, the nymphs of periodical cicadas live underground, sucking juices from the roots of forest trees all over the eastern half of the United States (except for our southern states, where a very similar or identical group of species emerges every 13 years). Then, within just a few weeks, millions of mature nymphs emerge from the ground, become adults, mate, lay their eggs, and die. (The best accounts, from an evolutionary standpoint, will be found in a series of articles by M. Lloyd and H. S. Dybas, published in the journals *Evolution* in 1966 and *Ecological Monographs* in 1974). Most remarkable is the fact that not one, but three separate species of periodical cicadas follow precisely the same schedule, emerging together in strict synchrony. Different areas may be out of phase—populations around Chicago do not emerge in the same year as forms from New England. But the 17-year cycle (13 years in the south) is invariant for each "brood"—the three species always emerge together in the same place. Janzen recognizes that cicadas and bamboo, despite their biological and geographic distance, represent the same evolutionary problem. Recent studies, he writes, "reveal no conspicuous qualitative difference between these insects and bamboo except perhaps in the way they count years."

As evolutionists, we seek answers to the question "why." Why, in particular, should such striking synchroneity evolve, and why should the period between episodes of sexual reproduction be so long? As I argued in discussing the matricidal habits of certain flies (essay 10) the theory of natural selection receives its strongest support when we devise satisfactory explanations for phenomena that strike us intuitively as bizarre or senseless.

In this case, we are confronted with a problem beyond the apparent peculiarity of such wastefulness (for very few seeds can sprout upon such saturated ground). The synchroneity of flowering or emergence seems to reflect an ordering and harmony operating upon the species as a whole, not upon its individual members. Yet Darwinian theory advocates no higher principle beyond individuals pursuing their own self-interest—i.e. the representation of their own genes in future generations. We must ask what advantage the synchroneity of sex provides for an individual cicada or bamboo plant.

The problem is similar to that faced by Adam Smith when he advocated an unbridled policy of laissez faire as the surest path to a harmonious economy. The ideal economy, Smith argued, might appear orderly and well balanced, but it would emerge "naturally" from the interplay of individuals who follow no path beyond the pursuit of their own best interests. The apparent direction towards a higher harmony, Smith argues in his famous metaphor, only reflects the operation of an "invisible hand."

> As every individual . . . by directing (his) industry in such a manner as its produce may be of greatest value, intends only his own gain, he is in this as in many other cases led by an invisible hand to promote an end which was no part of his intention. . . . By pursuing his own interest he frequently promotes that of society more effectively than when he really intends to promote it.

Since Darwin grafted Adam Smith upon nature to establish his theory of natural selection, we must seek an explanation for apparent harmony in the advantage that it confers upon individuals. What, then, does an individual cicada or bamboo gain by indulging in sex so rarely and at the same time as all its compatriots?

In order to appreciate the most likely explanation, we must recognize that human biology often provides a poor model for the struggles of other organisms. Humans are slowly growing animals. We invest a great deal of energy in raising very few, late maturing offspring. Our populations are not controlled by the wholesale death of nearly all juvenile members. Yet many organisms follow a different strategy in the "struggle for existence": they produce vast numbers of seeds or eggs, hoping (so to speak) that a few will survive the rigors of early life. These organisms are often controlled by their predators, and their evolutionary defense must be a strategy that minimizes the chance of being eaten. Cicadas and bamboo seeds seem to be particularly tasty to a wide variety of organisms.

Natural history, to a large extent, is a tale of different adaptations to avoid predation. Some individuals hide, others taste bad, others grow spines or thick shells, still others evolve to look conspicuously like a noxious relative; the list is nearly endless, a stunning tribute to nature's variety. Bamboo seeds and cicadas follow an uncommon strategy: they are eminently and conspicuously available, but so rarely and in such great numbers that predators cannot possibly consume the entire bounty. Among evolutionary biologists, this defense goes by the name of "predator satiation."

An effective strategy of predator satiation involves two adaptations. First, the synchrony of emergence or reproduction must be very precise, thus assuring that the market is truly flooded, and only for a short time. Secondly, this flooding cannot occur very often, lest predators simply adjust their own life cycle to predictable times of superfluity. If bamboos flowered every year, seed eaters would track the cycle and present their own abundant young with the annual bounty. But if the period between episodes of flowering far exceeds the life-span of any predator, then the cycle cannot be tracked (except by one peculiar primate that records its own history). The advantage of synchroneity to individual bamboos and cicadas is clear enough: anyone out of step is quickly gobbled up (cicada "stragglers" do occasionally emerge in off years, but they never gain a foothold).

The hypothesis of predator satiation, though unproven, meets the primary criterion of a successful explanation: it coordinates a suite of observations that would otherwise remain unconnected and, in this case, downright peculiar. We know, for example, that bamboo seeds are relished by a wide variety of animals, including many vertebrates with long life spans; the rarity of flowering cycles shorter than 15 or 20 years makes sense in this context. We also know that the synchronous setting of seed can inundate an affected area. Janzen records a mat of seeds 6 inches deep below the parental plant in one case. Two species of Malagasy bamboos produced 50 kilograms of seed per hectare over a large area of

100,000 hectares during a mass flowering.

The synchrony of three species among cicadas is particularly impressive—especially since years of emergence vary from place to place, while all three species invariably emerge together in any one area. But I am most impressed by the timing of the cycles themselves. Why do we have 13 and 17 year cicadas, but no cycles of 12, 14, 15, 16, or 18? 13 and 17 share a common property. They are large enough to exceed the life cycle of any predator, but they are also prime numbers (divisible by no integer smaller than themselves). Many potential predators have 2–5-year life cycles. Such cycles are not set by the availability of periodical cicadas (for they peak too often in years of nonemergence), but cicadas might be eagerly harvested when the cycles coincide. Consider a predator with a cycle of five years: if cicadas emerged every 15 years, each bloom would be hit by the predator. By cycling at a large prime number, cicadas minimize the number of coincidences (every 5 × 17, or 85 years, in this case). Thirteen- and 17-year cycles cannot be tracked by any smaller number.

Existence is, as Darwin stated, a struggle for most creatures. The weapons of survival need not be claws and teeth; patterns of reproduction may serve as well. Occasional superfluity is one pathway to success. It is sometimes advantageous to put all your eggs in one basket—but be sure to make enough of them, and don't do it too often.

12 The Problem of Perfection, or How Can a Clam Mount a Fish on Its Rear End?

IN 1802, Archdeacon Paley set out to glorify God by illustrating the exquisite adaptation of organisms to their appointed roles. The mechanical perfection of the vertebrate eye inspired a rapturous discourse on divine benevolence; the uncanny similarity of certain insects to pieces of dung also excited his admiration, for God must protect all his creatures, great and small. Evolutionary theory eventually unraveled the archdeacon's grand design, but threads of his natural theology survive.

Modern evolutionists cite the same plays and players; only the rules have changed. We are now told, with equal wonder and admiration, that natural selection is the agent of exquisite design. As an intellectual descendant of Darwin, I do not doubt this attribution. But my confidence in the power of natural selection has other roots: it is not based upon "organs of extreme perfection and complication," as Darwin called them. In fact, Darwin saw truly exquisite design as a problem for his theory. He wrote:

> To suppose that the eye with all its inimitable contrivances for adjusting the focus to different distances, for admitting different amounts of light, and for the correction of spherical and chromatic aberration, could have been formed by natural selection, seems, I confess, absurd in the highest degree.

In essay 10, I invoked gall midges to illustrate the opposite problem of adaptation—structures and behaviors that seem senseless. But "organs of extreme perfection" proclaim their value unambiguously; the difficulty lies in explaining how they developed. In Darwinian theory, complex adaptations do not arise in a single step, for natural selection would then be confined to the purely destructive task of eliminating the unfit whenever a better-adapted creature suddenly appeared. Natural selection has a constructive role in Darwin's system: it builds adaptation gradually, through a sequence of intermediate stages, by bringing together in sequential fashion elements that seem to have meaning only as parts of a final product. But how can a series of reasonable intermediate forms be constructed? Of what value could the first tiny step toward an eye be to its possessor? The dung-mimicking insect is well protected, but can there be any edge in looking only 5 percent like a turd? Darwin's critics referred to this dilemma as the problem of assigning adaptive value to "incipient stages of useful structures." And Darwin rebutted by trying to find the intermediate stages and by specifying their utility.

> Reason tells me, that if numerous gradations from a simple and imperfect eye to one complex and perfect can be shown to exist, each grade being useful to its possessor . . . then the difficulty of believing that a perfect and complex eye could be formed by natural selection, though insuperable by our imagination, should not be considered as subversive of the theory.

The argument still rages, and organs of extreme perfection rank high in the arsenal of modern creationists.

Every naturalist has his favorite example of an awe-inspiring adaptation. Mine is the "fish" found in several species of the freshwater mussel *Lampsilis.* Like most clams, *Lampsilis* lives partly buried in bottom sediments, with its posterior end protruding. Riding atop the protruding end is a structure that looks for all the world like a little fish. It has a streamlined body, well-designed side flaps complete with a tail and even an eyespot. And, believe it or not, the flaps undulate with a rhythmic motion that imitates swimming.

Most clams release their eggs directly into the surrounding

"Fish" with eyespot and tail rides atop **Lampsilis ventricosa.** *When a fish nears, the clam discharges larvae; some will be ingested by the fish and find their way to its gills, where they will mature.* (John H. Welsh)

water, where they are fertilized and undergo their embryonic development. But female unionids (the technical name for freshwater mussels) retain their eggs within their bodies, where they are fertilized by sperm released into the water by nearby males. The fertilized eggs develop in tubes within the gills, forming a brood pouch, or marsupium.

In *Lampsilis,* the inflated marsupium of gravid females forms the "body" of its ersatz fish. Surrounding the fish, symmetrically on both sides, are extensions of the mantle, the "skin" that encloses the soft parts of all clams and usually ends at the shell margin. These extensions are elaborately shaped and colored to resemble a fish, with a definite, often flaring "tail" at one end and an "eyespot" at the other. A special ganglion located inside the mantle edge innervates these flaps. As the flaps move rhythmically, a pulse, beginning at the tail, moves slowly forward to propel a bulge in the flaps along the entire body. This intricate apparatus, formed by the marsupium and mantle flaps, not only looks like a fish but also moves like one.

Why would a clam mount a fish on its rear end? The unusual reproductive biology of *Lampsilis* supplies an answer. The larvae of unionids cannot develop without a free ride upon fishes during their early growth. Most unionid larvae possess two little hooks. When released from their mother's marsupium, they fall to the bottom of the stream and await a passing fish. But the larvae of *Lampsilis* lack these hooks and cannot actively attach themselves. In order to survive, they must enter a fish's mouth and move to favored sites on the gills. The ersatz fish of *Lampsilis* is an animated decoy, simulating both the form and movement of the animal it must attract. When a fish approaches, *Lampsilis* discharges larvae

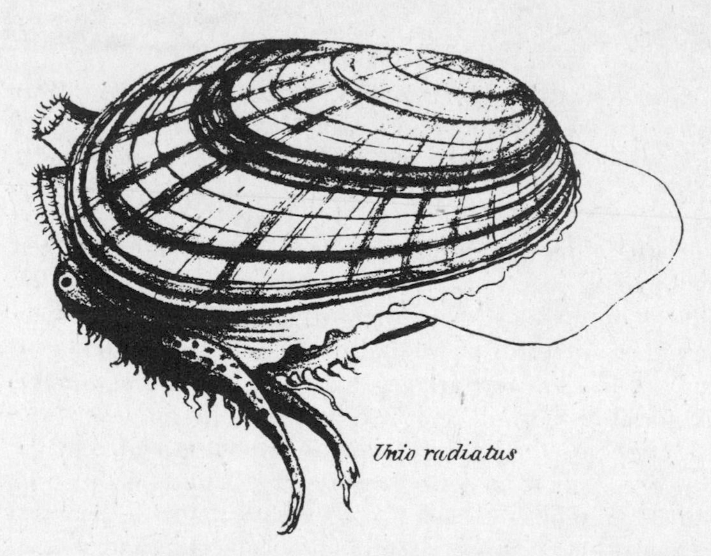

Isaac Lea published this figure of the decoy "fish" in 1838. I thank John H. Welsh for sending this figure to me.

from the marsupium; some of them will be swallowed by the fish and find their way to its gills.

The strategem of *Cyprogenia,* a related genus, emphasizes the importance of attracting a host. These mussels "go fishing" in a manner subsequently reinvented by disciples of Izaak Walton. The larvae attach themselves to a bright red "worm" formed by a protein manufactured within the mother's body. The "worms" are extruded through the exhalant siphon. Several observers report that fish seek out and eat these "worms," often pulling them, when only partly extruded, from the female's siphon.

We can scarcely doubt the adaptive significance of the decoy "fish," but how could it ever evolve? How did the marsupium and mantle flap come together to effect their ruse? Lucky accident or preordained direction may appeal more to our intuition than gradual construction by natural selection through some intermediate forms that, at least in their initial stages, could not have looked much like a fish. The intricate fish of *Lampsilis* is a classic illustration of a deep dilemma in Darwinism. Can we possibly devise an adaptive significance for the incipient stages of this useful structure?

The general principle advanced by modern evolutionists to solve this dilemma calls upon a concept with the unfortunate name of "preadaptation." (I say unfortunate because the term implies that species adapt in advance to impending events in their evolutionary history, when exactly the opposite meaning is intended.) The success of a scientific hypothesis often involves an element of surprise. Solutions often arise from a subtle reformulation of the question, not from the diligent collection of new information in an old framework. With preadaptation, we cut through the dilemma of a function for incipient stages by accepting the standard objection and admitting that intermediate forms did not work in the same way as their perfected descendants. We avoid the excellent question, What good is 5 percent of an eye? by arguing that the possessor of such an incipient structure did not use it for sight.

To invoke a standard example, the first fishes did not have jaws. How could such an intricate device, consisting of several interlocking bones, ever evolve from scratch? "From scratch" turns out to be a red herring. The bones were present in ancestors, but they were doing something else—they were supporting a gill arch located just behind the mouth. They were well designed for their respiratory role; they had been selected for this alone and "knew" nothing of any future function. In hindsight, the bones were admirably preadapted to become jaws. The intricate device was already assembled, but it was being used for breathing, not eating.

Similarly, how could a fish's fin ever become a terrestrial limb? Most fishes build their fins from slender parallel rays that could not support an animal's weight on land. But one peculiar group of freshwater, bottom-dwelling fishes—our ancestors—evolved a fin with a strong central axis and only a few radiating projections. It was admirably preadapted to become a terrestrial leg, but it had evolved purely for its own purposes in water—presumably for scuttling along the bottom by sharp rotation of the central axis against the substrate.

In short, the principle of preadaptation simply asserts that a structure can change its function radically without altering its form as much. We can bridge the limbo of intermediate stages by arguing for a retention of old functions while new ones are developing.

Will preadaptation help us to understand how *Lampsilis* got its fish? It might if we can meet two conditions: (1) We must find an intermediate form using at least some elements of the fish for different purposes; (2) We must specify functions other than visual decoy that the proto-fish could fulfill while it gradually acquired its uncanny resemblance.

Ligumia nasuta, a "cousin" of *Lampsilis,* seems to satisfy the first condition. Gravid females of this species do not have mantle flaps, but they do possess darkly pigmented, ribbon-like membranes that bridge the gap between partly opened shells. *Ligumia* uses these membranes to produce an unusual, rhythmic motion. The opposing edges of the ribbons part to form a gap several millimeters in length at the mid-part of the shell. Through this gap, the white color of the interior soft parts stands out against the dark pigment of the ribbon. This white spot appears to move toward the back of the shell, as a wave of separation propagates itself along the membranes. These waves may repeat about once every two seconds. J.H. Welsh wrote in the May 1969 issue of *Natural History:*

> The regularity of the rhythm is remarkably constant. To a human observer, and perhaps to a fish, the eye-catching feature here is the white spot that appears to move against the dark background of the mussel and the substrate in which it is half buried. Certainly this could be a lure to host fish and may represent a specialized adaptation from which the more elaborate, fishlike lure evolved.

We are still dealing with a device to attract fish, but the mechanism is abstract, regular motion, not visual mimicry. If this device operated while the flaps were evolving and slowly building their resemblance to a fish, then we have no problem of incipient stages. Motion of the mantle attracted fish from the start; the slow development of an "alternate technology" only enhanced the process.

Lampsilis itself fulfills the second condition. Although no one has denied the significance of visual resemblance as a lure, our leading student of *Lampsilis,* L.R. Kraemer, questions the common assumption that "flapping" of the body serves only to simulate the movements of a fish. She believes that flapping may have evolved either to aerate the larvae within the marsupium or to keep them suspended in the water after their release. Again, if flapping provided these other advantages from the start, then the fortuitous resemblance of flaps to fish might be a preadaptation. The initial, imperfect mimicry could be improved by natural selection while the flaps performed other important functions.

Common sense is a very poor guide to scientific insight for it represents cultural prejudice more often than it reflects the native honesty of a small boy before the naked emperor. Common sense dictated to Darwin's critics that a gradual change in form must indicate a progressive building of function. Since they could assign no adaptive value to early and imperfect stages of a function, they assumed either that early stages had never existed (and that perfect forms had been created all at once) or that they had not arisen by natural selection. The principle of preadaptation—functional change in structural continuity—can resolve this dilemma. Darwin ended his paragraph on the eye with this perceptive evaluation of "common sense":

> When it was first said that the sun stood still and the world turned round, the common sense of mankind declared the doctrine false; but the old saying of *Vox populi, vox Dei* [the voice of the people is the voice of God], as every philosopher knows, cannot be trusted in science.

4 Patterns and Punctuations in the History of Life

13 The Pentagon of Life

WHEN I WAS 10 years old, James Arness terrified me as a giant, predaceous carrot in *The Thing* (1951). A few months ago, older, wiser, and somewhat bored, I watched its latest television rerun with a dominating sentiment of anger. I recognized the film as a political document, expressing the worst sentiments of America in the cold war: its hero, a tough military man who wants only to destroy the enemy utterly; its villain, a naively liberal scientist who wants to learn more about it; the carrot and its flying saucer, a certain surrogate for the red menace; the film's famous last words—a newsman's impassioned plea to "watch the skies" —an invitation to extended fear and jingoism.

Amidst all this, a scientific thought crept in by analogy and this essay was born—the fuzziness of all supposedly absolute taxonomic distinctions. The world, we are told, is inhabited by animals with conceptual language (us) and those without (everyone else). But chimps are now talking (see essay 5). All creatures are either plants or animals, but Mr. Arness looked rather human (if horrifying) in his role as a mobile, giant vegetable.

Either plants or animals. Our basic conception of life's diversity is based upon this division. Yet it represents little more than a prejudice spawned by our status as large, terrestrial animals. True, the macroscopic organisms surrounding us on land can be unambiguously allocated if we designate fungi as plants because they are rooted (even though they do not photosynthesize). Yet, if we floated as tiny creatures in the oceanic plankton, we would not have made such a distinction. At the one-celled level, ambiguity abounds: mobile "animals" with functioning chloroplasts; simple cells like bacteria with no clear relation to either group.

Taxonomists have codified our prejudice by recognizing just two kingdoms for all life—Plantae and Animalia. Readers may regard an inadequate classification as a trifling matter; after all, if we characterize organisms accurately, who cares if our basic categories do not express the richness and complexity of life very well? But a classification is not a neutral hat rack; it expresses a theory of relationships that controls our concepts. The Procrustean system of plants and animals has distorted our view of life and prevented us from understanding some major features of its history.

Several years ago, Cornell ecologist R. H. Whittaker proposed a five-kingdom system for the organization of life (*Science,* January 10, 1969); his scheme has recently been championed and expanded by Boston University biologist Lynn Margulis (*Evolutionary Biology,* 1974). Their criticism of the traditional dichotomy begins among the single-celled creatures.

Anthropocentrism has a remarkably broad range of consequences, ranging from strip mining to whale killing. In folk taxonomy it merely leads us to make fine distinctions among creatures close to us and very broad ones for more distant, "simple" organisms. Every novel bump on a tooth defines a new kind of mammal, but we tend to lump all single-celled creatures together as "primitive" organisms. Nonetheless, specialists are now arguing that the most fundamental distinction among living things is not between "higher" plants and animals; it is a division *within* single-celled creatures—bacteria and blue-green algae on the one side, other groups of algae and protozoans (amoebae, paramecia, and so on) on the other. And neither group, according to Whittaker and Margulis, can be fairly called either plant or animal; we must have two new kingdoms for single-celled organisms.

Bacteria and blue-green algae lack the internal structures, or "organelles," of higher cells. They have no nucleus, chromosomes, chloroplasts, or mitochondria (the "energy factories" of higher cells). Such simple cells are called "prokaryotic" (roughly, before nuclei, from the Greek *karyon,* meaning "kernel"). Cells with organelles are termed "eukaryotic" (truly nucleate). Whittaker considers this distinction "the clearest, most effectively discontinuous separation of levels of organization in the living world." Three different arguments emphasize the division:

1. The history of prokaryotes. Our earliest evidence of life dates from rocks about three billion years old. From then until at least one billion years ago, all fossil evidence points to the existence of prokaryotic organisms only; for two billion years, blue-green algal mats were the most complicated forms of life on earth. Thereafter, opinion differs. UCLA paleobotanist J. W. Schopf believes that he has evidence for eukaryotic algae in Australian rocks about a billion years old. Others contend that Schopf's organelles are really the post-mortem degradation products of prokaryotic cells. If these critics are right, then we have no evidence for eukaryotes until the very latest Precambrian, just before the great Cambrian "explosion" of 600 million years ago (see essays 14 and 15). In any case, prokaryotic organisms held the earth as their exclusive domain during two-thirds to five-sixths of the history of life. With ample justice, Schopf labels the Precambrian as the "age of blue-green algae."

2. A theory for the origin of the eukaryotic cell. Margulis has stirred a great deal of interest in recent years with her modern defense of an old theory. The idea sounds patently absurd at first, but it quickly comes to compel attention, if not assent. I am certainly rooting for it. Margulis argues that the eukaryotic cell arose as a colony of prokaryotes—that, for example, our nucleus and mitochondria had their origins as independent prokaryotic organisms. Some modern prokaryotes can invade and live as symbionts within eukaryotic cells. Most prokaryotic cells are about the same size as eukaryotic organelles; the chloroplasts of photosynthetic eukaryotes are strikingly similar to the entire cells of some blue-green algae. Finally, some organelles have their own self-replicating genes, remnants of their formerly independent status as entire organisms.

3. The evolutionary significance of the eukaryotic cell.

Advocates of contraception have biology firmly on their side in arguing that sex and reproduction serve different purposes. Reproduction propagates a species, and no method is more efficient than the asexual budding and fission employed by prokaryotes. The biological function of sex, on the other hand, is to promote variability by mixing the genes of two (or more) individuals. (Sex is usually combined with reproduction because it is expedient to do the mixing in an offspring.)

Major evolutionary change cannot occur unless organisms maintain a large store of genetic variability. The creative process of natural selection works by preserving favorable genetic variants from an extensive pool. Sex can provide variation on this scale, but efficient sexual reproduction requires the packaging of genetic material into discrete units (chromosomes). Thus, in eukaryotes, sex cells have half the chromosomes of normal body cells. When two sex cells join to produce an offspring, the original amount of genetic material is restored. Prokaryotic sex, on the other hand, is infrequent and inefficient. (It is unidirectional, involving the transfer of a few genes from a donor cell to a recipient.)

Asexual reproduction makes identical copies of parental cells, unless a new mutation intervenes to yield a minor change. But new mutation is infrequent and asexual species do not maintain enough variability for significant evolutionary change. For two billion years, algal mats remained algal mats. But the eukaryotic cell made sex a reality; and, less than a billion years later, here we are—people, cockroaches, seahorses, petunias, and quahogs.

We should, in short, use the highest taxonomic distinction available to recognize the difference between prokaryotic and eukaryotic single-celled organisms. This establishes two kingdoms among one-celled creatures: Monera for the prokaryotes (bacteria and blue-green algae); Protista for the eukaryotes.

Among multicellular organisms, Plantae and Animalia remain in their traditional senses. Whence, then, the fifth kingdom? Consider the fungi. Our Procrustean dichotomy forced them into Plantae, presumably because they are rooted to a single spot. But their resemblance to true plants stops with this misleading feature. Higher fungi maintain a system of tubes superficially like those of plants; but while nutrients flow in plants, protoplasm itself courses through the fungal tubes. Many fungi reproduce by combining the nuclei of several individuals into a multinucleate tissue without nuclear fusion. The list could be extended, but all its items pale before one cardinal fact: fungi do not photosynthesize. They live embedded in their food source and feed by absorption (often by excreting enzymes for external digestion). Fungi, then, form the fifth and final kingdom.

As Whittaker argues, the three kingdoms of multicellular life represent an ecological, as well as a morphological, classification. The three major ways of making a living in our world are well represented by plants (production), fungi (reduction), and animals (consumption). And, as another nail in the coffin of our self-importance, I hasten to point out that the major cycle of life runs between production and reduction. The world could get along very well without its consumers.

I like the five-kingdom system because it tells a sensible story about organic diversity. It arranges life in three levels of increasing complexity: the prokaryotic unicells (Monera), the eukaryotic unicells (Protista), and the eukaryotic multicells (Plantae, Fungi, and Animalia). Moreover, as we ascend through the levels, life becomes more diverse—as we should expect since increasing complexity of design begets more opportunity for variation upon it. The world contains more distinctively different kinds of protistans than monerans. At the third level, diversity is so great that we need three separate kingdoms to encompass it. Finally, I note that the evolutionary transition from any level to the next occurs more than once; the advantages of increased complexity are so great that many independent lines converge upon the few possible solutions. The members of each kingdom are united by common structure, not by common descent. In Whittaker's view, plants evolved at least four separate times from protistan ancestors, fungi at least five times, and animals at least three times (the peculiar mesozoans, sponges, and everything else).

The three-leveled, five-kingdom system may appear, at first glance, to record an inevitable progress in the history of life. Increasing diversity and multiple transitions seem to reflect a determined and inexorable progression toward higher things. But the paleontological record supports no such interpretation. There has been no steady progress in the higher development of organic design. We have had, instead, vast stretches of little or no change and one evolutionary burst that created the entire system. For the first two-thirds to five-sixths of life's history, monerans alone inhabited the earth, and we detect no steady progress from "lower" to "higher" prokaryotes. Likewise, there has been no addition of basic designs since the Cambrian explosion filled our biosphere (although we can argue for limited improvement *within* a few designs—vertebrates and vascular plants, for example).

Rather, the entire system of life arose during about 10 percent of its history surrounding the Cambrian explosion some 600 million years ago. I would identify two main events: the evolution of the eukaryotic cell (making further complexity possible by providing genetic variability through efficient sexual reproduction) and the filling of the ecological barrel by an explosive radiation of multicellular eukaryotes.

The world of life was quiet before and it has been relatively quiet ever since. The recent evolution of consciousness must be viewed as the most cataclysmic happening since the Cambrian if only for its geologic and ecological effects. Major events in evolution do not require the origin of new designs. The flexible eukaryotes will continue to produce novelty and diversity so long as one of its latest products controls itself well enough to assure the world a future.

14 An Unsung Single-Celled Hero

ERNST HAECKEL, the great popularizer of evolutionary theory in Germany, loved to coin words. The vast majority of his creations died with him a half-century ago, but among the survivors are "ontogeny," "phylogeny," and "ecology." The last is now facing an opposite fate—loss of meaning by extension and vastly inflated currency. Common usage now threatens to make "ecology" a label for anything good that happens far from cities or anything that does not have synthetic chemicals in it. In its more restricted and technical sense, ecology is the study of organic diversity. It focuses on the interactions of organisms and their environments in order to address what may be the most fundamental question in evolutionary biology: "Why are there so many kinds of living things?"

During the first century of Darwinism, ecologists pursued this question with little success. In the face of life's overwhelming complexity, they chose the empirical route and amassed storehouses of data on simple systems in limited areas. Now, nearly twenty years after the centennial of Darwin's *Origin of Species,* this poor sister among evolutionary disciplines has become a leader. Spurred by the efforts of scientists with a mathematical bent, ecologists have built theoretical models of organic interaction and applied them successfully to explain data from the field. We are finally beginning to understand (and quantify) the causes of organic diversity.

An important scientific advance usually extends its influence by providing keys to the resolution of persistent problems in related fields. Theoretical ecology, which works in the smallest dimension of "ecological" time (organic interactions over seasons or, at most, years), has begun to influence paleontology, the custodian of the longest dimension of all —three billion years of the history of life. In essay 16, I discuss how an ecological theory relating organic diversity to habitable area may have solved the great mystery of the Permian extinction. Here I will argue that another ecological theory, the relationship of diversity to predation, may provide a major clue in solving the second greatest dilemma of paleontology: the Cambrian "explosion" of life.

About 600 million years ago, at the beginning of what geologists call the Cambrian period, most of the major phyla of invertebrate animals made their appearance within the short span of a few million years. What had happened during the previous four billion years of earth history? What was it about the early Cambrian world that could have inspired such a burst of evolutionary activity?

These questions have disturbed paleontologists ever since the evolutionary view triumphed more than a century ago. For although rapid bursts of evolution and massive waves of extinction are not inconsistent with Darwinian theory, a deeply rooted bias of Western thought predisposes us to look for continuity and gradual change: *natura non facit saltum* ("nature does not make leaps"), as the older naturalists proclaimed.

The Cambrian explosion so disturbed Charles Darwin that he wrote in the last edition of his *Origin of Species:* "The case at present must remain inexplicable; and may be truly urged as a valid argument against the views here entertained." The situation was, indeed, far worse in Darwin's day. At that time, not a single Precambrian fossil had been found, and the Cambrian explosion of complex invertebrates provided the earliest evidence for any life on earth. If so many forms of life arose at the same time and with such initial complexity, might one not argue that God had chosen the base of the Cambrian for His moment (or six days) of creation?

Darwin's difficulty has been partly circumvented. We now have records of Precambrian life stretching back more than three billion years. Fossil bacteria and blue-green algae have been recovered in several places from rocks dated at between two and three billion years in age.

Nonetheless, these exciting finds in Precambrian paleontology do not remove the problem of the Cambrian explosion, for they include only the simple bacteria and blue-green algae (see essay 13), and some higher plants such as green algae. The evolution of complex Metazoa (multicelled animals) seems as sudden as ever. (A single Precambrian fauna has been found at Ediacara in Australia. It includes some relatives of modern fan corals, jellyfish, wormlike creatures, arthropods, and two cryptic forms unlike anything alive today. Yet the Ediacara rocks lie just below the base of the Cambrian and qualify as Precambrian only by the slimmest margin. A few more isolated finds from other areas around the world are likewise just barely Precambrian.) If anything, the problem is increased because exhaustive study of more and more Precambrian rocks destroys the old and popular argument that complex Metazoa are really there, but we just haven't found them yet.

The last century of argument has produced only two basic strategies for a scientific explanation of the Cambrian explosion.

First, we may argue that it is a false appearance. Evolution was really slow and gradual, as Western biases dictate. The so-called explosion only marks the first appearance in the fossil record of creatures that had been living and developing for a long part of the Precambrian. But what prevented the fossilization of such rich faunas? Here we have a variety of proposals ranging from the absurdly *ad hoc* to the eminently plausible. To cite just a few:

(1) The Cambrian represents the first preservation of unaltered rocks; Precambrian sediments have been subjected to such heat and pressure that their fossil remains have been obliterated. This is empirically false, beyond any doubt.

(2) Life evolved in terrestrial lakes. The Cambrian represents the migration of this fauna to the sea.

(3) All early metazoans were soft-bodied. The Cambrian represents the evolution of fossilizable hard parts.

The popularity of this first strategy has plummeted with the discovery of abundant Precambrian fossil deposits devoid of anything more complex than algae. Nonetheless, the argument based on hard parts probably contains an element of truth, though it cannot provide the entire answer. A clam without a shell is not a viable animal; you cannot clothe any simple soft-bodied organism to make one. The delicate gills and the complex musculature clearly evolved in association with a hard outer covering. Hard parts often require a simultaneous and complex modification of any conceivable soft-bodied ancestor; their sudden appearance in the Cambrian, therefore, implies a truly rapid evolution of the animal they cover.

As a second strategy, we may claim that the Cambrian explosion is a real event representing the extremely rapid evolution of complexity. Something must have happened to the environment of simple, soft-bodied precursors of Cambrian metazoans in order to engender such a rapid burst of evolution. We have only two overlapping possibilities: changes in the physical or in the biological environment.

In 1965, Lloyd V. Berkner and Lauriston C. Marshall, two physical scientists from Dallas, published a famous article proposing that levels of oxygen in the earth's atmosphere exerted a direct physical control on the Cambrian explosion of life. Geologists agree that the earth's original atmosphere contained little or no free oxygen. Oxygen built up gradually as a result of organic activity—the photosynthesis of Precambrian algae. Metazoans require high levels of free oxygen for two reasons: directly, for respiration; indirectly because oxygen, in the form of ozone, absorbs harmful ultraviolet radiation in the upper atmosphere before it reaches life on the earth's surface. Berkner and Marshall simply proposed that the base of the Cambrian marks the first time that atmospheric oxygen reached a level sufficient for respiration and the shielding of harmful radiation.

But this attractive notion has foundered on the geologic evidence. Photosynthesizing organisms were probably abundant more than two and a half billion years ago. Is it reasonable to suppose that some two billion years were required for the buildup of sufficient oxygen for respiration? Moreover, many extensive deposits between one and two billion years old contain large volumes of strongly oxidized rocks.

Berkner and Marshall's hypothesis embodies an attitude all too common among nonbiologists who lack sufficient appreciation for the complexity that makes a machine a poor model for a living organism. Physical models often employ simple, inert objects like billiard balls that respond automatically to the impress of physical forces. But an organism cannot be pushed around so easily; it certainly does not evolve automatically. Berkner and Marshall's hypothesis relies upon the billiard-ball thinking that I term "physicalism"—metazoans arise immediately and automatically when a physical barrier to their existence is removed. The presence of sufficient oxygen, however, does not guarantee the immediate evolution of everything that could breathe it. Oxygen is a necessary but woefully insufficient requirement for the evolution of metazoans. In fact, enough oxygen probably existed for a billion years before the Cambrian explosion. Perhaps we should look to biological controls.

Steven M. Stanley of Johns Hopkins University has recently argued that a popular ecological theory—the "cropping principle"—may provide such a biological control *(Proceedings of the National Academy of Sciences, 1973).* The great geologist Charles Lyell argued that a scientific hypothesis is elegant and exciting insofar as it contradicts common sense. The cropping principle is just such a counterintuitive notion. In considering the causes of organic diversity, we might expect that the introduction of a "cropper" (either a herbivore or a carnivore) would reduce the number of species present in a given area: after all, if an animal is cropping food from a previously virgin area, it ought to reduce diversity and remove completely some of the rarer species.

In fact, a study of how organisms are distributed yields the opposite expectation. In communities of primary producers (organisms that manufacture their own nutrients by photosynthesis and do not feed upon other creatures), one or a very few species will be superior in competition and will monopolize space. Such communities may have an enormous biomass, but they are usually impoverished in numbers of species. Now, a cropper in such a system tends to prey on the abundant species, thus limiting their ability to dominate and freeing space for other species. A well-evolved cropper decimates—but does not destroy—its favorite prey species (lest it eat itself to eventual starvation). A well-cropped ecosystem is maximally diverse, with many species and few individuals of any single species. Stated another way, the introduction of a new level in the ecological pyramid tends to broaden the level below it.

The cropping principle is supported by many field studies: predatory fish introduced in an artificial pond cause an increase in the diversity of zooplankton; removal of grazing sea urchins from a diverse algal community leads to the domination of that community by a single species.

Consider the Precambrian algal community that persisted for two and a half billion years. It consisted exclusively of simple, primary producers. It was uncropped and, for that reason, biologically monotonous. It evolved with exceeding slowness and never attained great diversity because its physical space was so strongly monopolized by a few abundant forms. The key to the Cambrian explosion, Stanley argues, is the evolution of cropping herbivores—single-celled protists that ate other cells. Croppers made space for a greater diversity of producers, and this increased diversity permitted the evolution of more specialized croppers. The ecological pyramid burst out in both directions, adding many species at lower levels of production and adding new levels of carnivory at the top.

How can one prove such a notion? The original cropping protist, perhaps the unsung hero of the history of life, probably was not fossilized. There is, however, some suggestive indirect evidence. The most abundant producer communities of the Precambrian are preserved as stromatolites (blue-green algal mats that trap and bind sediment). Today, stromatolites thrive only in hostile environments largely devoid of metazoan croppers (hypersaline lagoons, for example). Peter Garrett found that these mats persist in more normal marine environments only when croppers are artificially removed. Their Precambrian abundance probably reflects the absence of croppers.

Stanley did not develop his theory from empirical studies of Precambrian communities. It is a deductive argument based on an established principle of ecology that does not contradict any fact of the Precambrian world and seems particulary consistent with a few observations. In a frank concluding paragraph, Stanley presents four reasons for accepting his theory: (1) "It seems to account for what facts we have about Precambrian life"; (2) "It is simple, rather than complex or contrived"; (3) "It is purely biological, avoiding *ad hoc* invocation of external controls"; and (4) "It is largely the product of direct deduction from an established ecological principle."

Such justifications do not correspond to the simplistic notions about scientific progress that are taught in most high schools and advanced by most media. Stanley does not invoke proof by new information obtained from rigorous experiment. His second criterion is a methodological presumption, the third a philosophical preference, the fourth an application of prior theory. Only Stanley's first reason makes any reference to Precambrian facts, and it merely makes the weak point that his theory "accounts" for what is known (many other theories do the same).

But creative thought in science is exactly this—not a mechanical collection of facts and induction of theories, but a complex process involving intuition, bias, and insight from other fields. Science, at its best, interposes human judgment and ingenuity upon all its proceedings. It is, after all (although we sometimes forget it), practiced by human beings.

15 Is the Cambrian Explosion a Sigmoid Fraud?

RODERICK MURCHISON, urged on by his wife, gave up the joys of fox hunting for the more sublime pleasures of scientific research. This aristocratic geologist devoted much of his second career to documenting the early history of life. He discovered that the first stocking of the oceans did not occur gradually with the successive addition of ever more complex forms of life. Instead, most major groups seemed to arise simultaneously at what geologists now call the base of the Cambrian period some 600 million years ago. To Murchison, a devout creationist writing in the 1830s, this episode could only represent God's initial decision to populate the earth.

Charles Darwin viewed this observation with trepidation. He assumed, as evolution demanded, that the seas had "swarmed with living creatures" before the Cambrian period. To explain the absence of fossils in the earlier geologic record, he apologetically speculated that our modern continents accumulated no sediments during Precambrian times because they were covered by clear seas.

Our modern view synthesizes these two opinions. Darwin, of course, has been vindicated in his cardinal contention: Cambrian life did arise from organic antecedents, not from the hand of God. But Murchison's basic observation reflects a biological reality, not the imperfections of geologic evidence: the Precambrian fossil record is little more (save at its very end) than 2.5 billion years of bacteria and blue-green algae. Complex life did arise with startling speed near the base of the Cambrian. (Readers must remember that geologists have a peculiar view of rapidity. By vernacular standards, it is a slow fuse indeed that burns for 10 million years. Still, 10 million years is but 1/450 of the earth's history, a mere instant to a geologist.)

Paleontologists have spent a largely fruitless century trying to explain this Cambrian "explosion"—the steep rise in diversity during the first 10 to 20 million years of the Cambrian period. (see essay 14). They have assumed, universally, that the puzzling event is the explosion itself. Any adequate theory, therefore, would have to explain why the early Cambrian was such an unusual time: perhaps it represents the first accumulation of sufficient atmospheric oxygen for respiration, or the cooling down of an earth previously too hot to support complex life (simple algae survive at much higher temperatures than complex animals), or a change in oceanic chemistry permitting the deposition of calcium carbonate to clothe previously soft-bodied animals with preservable skeletons.

I now sense that a fundamental change in attitude is about to take hold within my profession. Perhaps we have been looking at this important problem the wrong way round. Perhaps the explosion itself was merely the predictable outcome of a process inexorably set in motion by an earlier Precambrian event. In such a case, we would not have to believe that early Cambrian times were "special" in any way; the cause of the explosion would be sought in an earlier event that initiated the evolution of complex life. I have recently been persuaded that this new perspective is probably correct. The pattern of the Cambrian explosion seems to follow a general law of growth. This law predicts a phase of steep acceleration; the explosion is no more fundamental (or in need of special explanation) than its antecedent period of slower growth or its subsequent leveling off. Whatever initiated the antecedent period virtually guaranteed the later explosion as well. In support of this new perspective, I offer two arguments based on a quantification of the fossil record. I hope not only to make my particular case but also to illustrate the role that quantification can play in testing hypotheses within professions that once eschewed such rigor.

The day-to-day work of field geology is a painstaking exercise in apparent minutiae of detail: the mapping of strata; their temporal correlation by fossils and by physical "superposition" (younger above older); the recording of rock types, grain sizes, and environments of deposition. This activity is often pooh-poohed by hotshot young theorists who regard it as the dog work of unimaginative drones. Yet we would have no science without the foundation that these data provide. In this case, our revised perspective on the Cambrian explosion rests upon a refinement of early Cambrian stratigraphy established primarily by Soviet geologists in recent years. The long Lower Cambrian has been subdivided into four stages and the first appearances of Cambrian fossils have been recorded with much greater accuracy. We can now tabulate a finely divided sequence of first appearances where previous stratigraphers could only record "Lower Cambrian" for all groups (thus accentuating the apparent explosion).

J. J. Sepkoski, a paleontologist at the University of Rochester, has recently found that a plot of increasing organic diversity versus time from the late Precambrian to the end of the "explosion" conforms to our most general model of growth—the so-called sigmoidal (S-shaped) curve. Consider the growth of a typical bacterial colony on a previously uninhabited medium: each cell divides every twenty minutes to form two daughters. Increase in population size is slow at first. (Rates of cell division are as fast as they will ever be, but founding cells are few in number and the population builds slowly toward its explosive period.) This "lag" phase forms the initial, slowly rising segment of the sigmoidal curve. The explosive, or "log," phase follows as each cell of a substantial population produces two fecund daughters every twenty minutes. Clearly this process cannot continue indefinitely: a not-too-distant extrapolation would fill the entire universe with bacteria. Eventually, the colony guarantees its own stability (or demise) by filling its space, exhausting its nutrients, fouling its nest with waste products, and so on. This leveling puts a ceiling on the log phase and completes the S of the sigmoidal distribution.

It is a long step from bacteria to the evolution of life, but sigmoidal growth is a general property of certain systems, and the analogy seems to hold in this case. For cell division, read speciation; for the agar substrate of a laboratory dish, read the oceans. The lag phase of life is the slow, initial rise of latest Precambrian times. (We now have a modest fauna of latest Precambrian age—mainly coelenterates [soft corals and jellyfish] and worms.) The famous Cambrian explosion is nothing more than the log phase of this continuous process, while post-Cambrian leveling represents the initial filling of ecological roles in the world's oceans (terrestrial life evolved later).

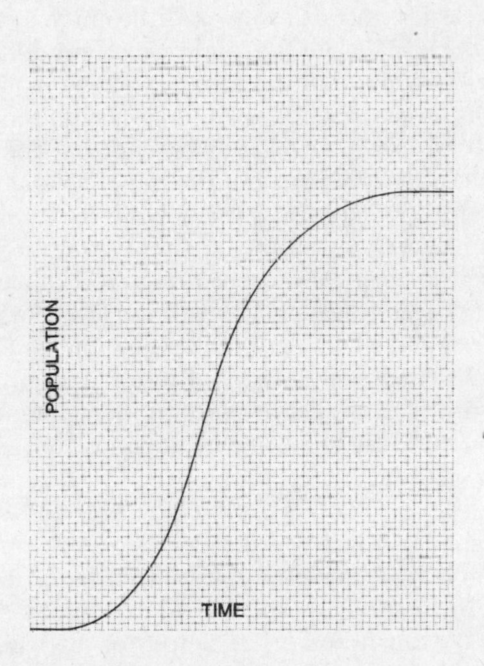

A typical sigmoidal (S-shaped) curve. Note slow beginning (lag phase), middle period of rapid increase (log phase) and final tapering off.

If the laws of sigmoidal growth regulated the early diversification of life, then there is nothing special about the Cambrian explosion. It is merely the log phase of a process determined by two factors: (1) the event that initiated the lag phase well within Precambrian times and (2) the properties of an environment that permitted sigmoidal growth.

As Johns Hopkins paleontologist S. M. Stanley wrote in a recent review (*American Journal of Science,* 1976): "We can abandon the traditional view that the origins of major fossil taxa near the start of the Cambrian . . . represent a major enigma. What remains as the 'Cambrian Problem' is the delay of the origin of multicellularity until the Earth was nearly 4 billion years old." We may deny the Cambrian problem by casting it back upon an earlier event, but the nature and cause of this earlier episode remains as the enigma of paleontological enigmas. The late Precambrian origin of the eukaryotic cell must provide an important determinant. (I argue in essay 13 that efficient sexual reproduction required a eukaryotic cell with discrete chromosomes, and that complex organisms could not evolve without the genetic variability that sexual reproduction supplies.) But we haven't the slightest idea why the eukaryotic cell arose when it did more than 2 billion years after the evolution of prokaryotic ancestors. In essay 14, I advocated Stanley's "cropping" theory for the initiation of sigmoidal increase following the evolution of eukaryotic cells. Stanley argues that Precambrian prokaryotic algae had usurped all available space in their potential habitat, thus precluding the evolution of anything more complex by denying a foothold to any competitor. The first eukaryotic herbivore, in the course of its copious, if unvaried, worldwide feast, freed enough space for the evolution of competitors.

Speculation may be intriguing, but we have little concrete to say about my first factor—the cause that initiated sigmoidal increase. We can, however, do better for the second—the nature of an environment that permitted it. Sigmoidal growth is not a universal property of natural systems; it occurs only in one kind of environment. Our laboratory bacteria would not have increased in an S-shaped curve if their medium had already been densely populated or devoid of nutrients. Sigmoidal patterns occur only in open, unconstrained systems, where food and space are so abundant that organisms grow until their own numbers limit further increase. The Precambrian oceans clearly formed such an "empty" ecosystem—plenty of space, abundant food, no competition. (The early eukaryotes could thank their prokaryotic ancestors not only for an immediate supply of food but also for their prior service in supplying oxygen to the atmosphere through photosynthesis.) The sigmoidal curve—with the Cambrian explosion as its log phase—represents the first stocking of the world's oceans, a predictable pattern of evolution in open ecosystems.

Animals evolving during the log phase should show different evolutionary patterns from those arising later in a regime of self-regulated equilibrium. Much of my own research in the past two years has been devoted to defining these differences. My colleagues (T. J. M. Schopf of the University of Chicago, D. M. Raup and J. J. Sepkoski of the University of Rochester, and D. S. Simberloff of Florida State University) and I have been modeling evolutionary trees as a random process. After "growing" a tree, we divide it into its major "limbs" and consider the history of each limb (technically called a "clade") through time. We depict each clade as a so-called spindle diagram. Spindle diagrams are constructed in the following way: simply count the number of species living during each period of time and vary the width of the diagram according to this number.

We then measure several properties of these diagrams. One measure, called C.G., defines the position of the center of gravity (roughly, the place where the clade is widest, or most diverse). If this position of maximum diversity occurs at the midpoint of the clade's duration, we give C.G. a value of 0.5 (halfway in the clade's total existence). If a clade reaches its greatest diversity before its midpoint, it has a C.G. of less than 0.5.

In our random system, C. G. is always near 0.5—the ideal clade is a diamond widest at its center. But our random world is one of perfect equilibrium. No log phases of sigmoidal growth are permitted; a constant number of species is maintained through time, as rates of extinction match rates of origination.

I spent a good part of 1975 counting fossil genera and recording their longevity in order to construct spindle diagrams for actual clades. I now have more than 400 clades for groups that arose and died *following* the log phase of the Cambrian explosion. Their mean value is 0.4993 —couldn't ask for anything closer to the 0.5 of our idealized world at equilibrium. I also have as many spindle diagrams for clades that arose *during* the log phase and died out afterward. Their mean C.G. is significantly less than 0.5. They record an atypical world of increasing diversity, and their values can be used to assess both the timing and the strength of the Cambrian log phase. Their values are below 0.5 because they arose during times of rapid diversification, but died out during

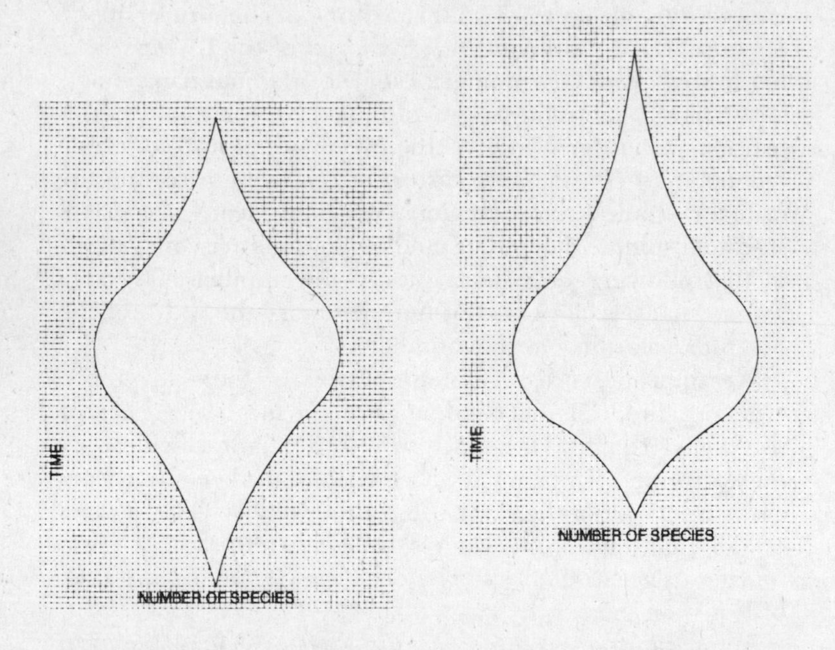

Spindle diagrams. The diagram on the left has a C.G. of 0.5 (widest at the midpoint of its duration); the diagram on the right has a C.G. lower than 0.5.

stable times of slower origin and extinction. Thus, they reached a maximum diversity early in their history since their first representatives participated in a log phase of unrestrained increase, but they petered out more slowly in the stabilized world that followed.

A quantitative approach has helped us to understand the Cambrian explosion in two ways. First, we can recognize its character of sigmoidal growth and identify its cause in an earlier event; the Cambrian problem, per se, disappears. Secondly, we can define the time and intensity of the Cambrian log phase by studying the statistics of spindle diagrams.

To my mind, the most remarkable result of this exercise is not the low C.G. of Cambrian clades, but the correspondence of C.G. for later clades to an idealized model for a world at equilibrium. Could it be that the diversity of marine life has remained at equilibrium through all the vicissitudes of an earth in motion, all the mass extinctions, the collision of continents, the swallowing up and creation of oceans? The log phase of the Cambrian filled up the earth's oceans. Since then, evolution has produced endless variation on a limited set of basic designs. Marine life has been copious in its variety, ingenious in its adaptation, and (if I may be permitted an anthropocentric comment) wondrous in its beauty. Yet, in an important sense, evolution since the Cambrian has only recycled the basic products of its own explosive phase.

16 | The Great Dying

ABOUT 225 MILLION years ago, at the end of the Permian period, fully half the families of marine organisms died out during the short span of a few million years—a prodigious amount of time by most standards, but merely minutes to a geologist. The victims of this mass extinction included all surviving trilobites, all ancient corals, all but one lineage of ammonites, and most bryozoans, brachiopods, and crinoids.

This great dying was the most profound of several mass extinctions that have punctuated the evolution of life during the past 600 million years. The late Cretaceous extinction, some 70 million years ago, takes second place. It destroyed 25 percent of all families, and cleared the earth of its dominant terrestrial animals, the dinosaurs and their kin—thus setting a stage for the dominance of mammals and the eventual evolution of man.

No problem in paleontology has attracted more attention or led to more frustration than the search for causes of these extinctions. The catalog of proposals would fill a Manhattan telephone book and include almost all imaginable causes: mountain building of world wide extent, shifts in sea level, subtraction of salt from the oceans, supernovae, vast influxes of cosmic radiation, pandemics, restriction of habitat, abrupt changes in climate, and so on. Nor has the problem escaped public notice. I remember well my first exposure to it at age five: the dinosaurs of Disney's *Fantasia* panting to their deaths across a desiccating landscape to the tune of Stravinsky's *Rite of Spring.*

Since the Permian extinction dwarfs all the others, it has long been the major focus of inquiry. If we could explain this greatest of all dyings, we might hold the key to understanding mass extinctions in general.

During the past decade, important advances in both geol-

The dinosaurs of Disney's "Fantasia" pant their way to extinction across a desiccating landscape. (© 1940 Walt Disney Productions)

ogy and evolutionary biology have combined to provide a probable answer. This solution has developed so gradually that some paleontologists scarcely realize that their oldest and deepest dilemma has been resolved.

Ten years ago, geologists generally believed that the continents formed where they now stand. Large blocks of land might move up and down and continents might "grow" by accretion of uplifted mountain chains at their borders, but continents did not wander about the earth's surface—their positions were fixed for all time. An alternative theory of continental drift had been proposed early in the century, but the absence of a mechanism for moving continents had assured its nearly universal rejection.

Now, studies of the ocean floor have yielded a mechanism in the theory of plate tectonics. The earth's surface is divided into a small number of plates bordered by ridges and subduction zones. New ocean floor is formed at the ridges as older portions of the plates are pushed away. To balance this addition, old parts of plates are drawn into the earth's interior at subduction zones.

Continents rest passively upon the plates and move with them; they do not "plow" through solid ocean floor as previous theories proposed. Continental drift, therefore, is but one consequence of plate tectonics. Other important consequences include earthquakes at plate boundaries (like the San Andreas Fault running past San Francisco) and mountain chains where two plates bearing continents collide (the Himalayas formed when the Indian "raft" hit Asia).

When we reconstruct the history of continental movements, we realize that a unique event occurred in the latest Permian: all the continents coalesced to form the single supercontinent of Pangaea. Quite simply, the consequences of this coalescence caused the great Permian extinction.

But which consequences and why? Such a fusion of fragments would produce a wide array of results, ranging from changes in weather and oceanic circulation to the interaction of previously isolated ecosystems. Here we must look to advances in evolutionary biology—to theoretical ecology and our new understanding of the diversity of living forms.

After several decades of highly descriptive and largely atheoretical work, the science of ecology has been enlivened by quantitative approaches that seek a general theory of organic diversity. We are gaining a better understanding of the influences of different environmental factors upon the abundance and distribution of life. Many studies now indicate that diversity—the numbers of different species present in a given area—is strongly influenced, if not largely controlled, by the amount of habitable area itself. If, for example, we count the number of ant species living on each of a group of islands differing only in size (and otherwise similar in such properties as climate, vegetation, and distance from the mainland), we find that, in general, the larger the island, the greater the number of species.

It is a long way from ants on tropical islands to the entire marine biota of the Permian period. Yet we have good reason to suspect that area might have played a major role in the great extinction. If we can estimate organic diversity and area for various times during the Permian (as the continents coalesced), then we can test the hypothesis of control by area.

We must first understand two things about the Permian extinction and the fossil record in general. First, the Permian extinction primarily affected marine organisms. The relatively few terrestrial plants and vertebrates then living were not so strongly disturbed. Second, the fossil record is very strongly biased toward the preservation of marine life in shallow water. We have almost no fossils of organisms inhabiting the ocean depths. Thus, if we want to test the theory that reduced area played a major role in the Permian extinction, we must look to the area occupied by shallow seas.

We can identify, in a qualitative way, two reasons why a coalescence of continents would drastically reduce the area of shallow seas. The first is basic geometry: If each separate land mass of pre-Permian times were completely surrounded by shallow seas, then their union would eliminate all area at the sutures. Make a single square out of four graham crackers and the total periphery is reduced by half. The second reason involves the mechanics of plate tectonics. When oceanic ridges are actively producing new sea floor to spread outward, then the ridges themselves stand high above the deepest parts of the ocean. This displaces water from the ocean basins, world sea level rises, and continents are partly flooded. Conversely, if spreading diminishes or stops, ridges begin to collapse and sea level falls.

When continents collided in the late Permian, the plates that carried them "locked" together. This set a brake upon new spreading. Ocean ridges sank and shallow seas withdrew from the continents. The drastic reduction in shallow seas was not caused by a drop in sea level *per se,* but rather by the configuration of sea floor over which the drop occurred. The ocean floor does not plunge uniformly from shoreline to ocean deep. Today's continents are generally bordered by a very wide continental shelf of persistently shallow water. Seaward of the shelf lies the continental slope of much greater steepness. If sea level fell far enough to expose the entire continental shelf, then most of the world's shallow seas would disappear. This may well have happened during the late Permian.

Thomas Schopf of the University of Chicago has recently tested this hypothesis of extinction by reduction in area. He studied the distribution of shallow water and terrestrial rocks to infer continental borders and extent of shallow seas for several times during the Permian as the continents coalesced. Then, by an exhaustive survey of the paleontological literature, he counted the numbers of different kinds of organisms living during each of these Permian times. Daniel Simberloff of Florida State University then showed that the standard mathematical equation relating numbers of species to area fits these data very well. Moreover, Schopf showed that the extinction did not affect certain groups differentially; its results were evenly spread over all shallow-water inhabitants. In other words, we do not need to seek a specific cause related to the peculiarities of a few animal groups. The effect was general. As shallow seas disappeared, the rich ecosystem of earlier Permian times simply lacked the space to support all its members. The bag became smaller and half the marbles had to be discarded.

Area alone is not the whole answer. Such a momentous event as the fusion of a single supercontinent must have entailed other consequences deterimental to the precariously balanced ecosystem of earlier Permian time. But Schopf and Simberloff have provided persuasive evidence for granting a major role to the basic factor of space.

It is gratifying that an answer to paleontology's outstanding dilemma has arisen as a by-product of exciting advances in two related disciplines—ecology and geology. When a problem has proved intractable for more than one hundred years, it is not likely to yield to more data collected in the old way and under the old rubric. Theoretical ecology allowed us to ask the right questions and plate tectonics provided the right earth upon which to pose them.

5 Theories of the Earth

17 The Reverend Thomas' Dirty Little Planet

"WE DO NOT seem to inhabit the same world that our first forefathers did. . . . To make one man easie, ten must work and do drudgery. . . . The earth doth not yield us food, but with much labor and industry. . . . The air is often impure or infectious."

Modern eco-activism this is not. The sentiment is right, but the style is a giveaway. It is, instead, the lament of Rev. Thomas Burnet, author of the most popular geologic work of the seventeenth century—*The Sacred Theory of the Earth.* His words depict a planet fallen from the original grace of Eden, not a world depleted by too many greedy men.

Among the works of scriptural geology, Burnet's *Sacred Theory* is surely the most famous, the most maligned, and the most misunderstood. In it, he tried to provide a geologic rationale for all biblical events, past and future. Now take a simplistic but common view of the relationship between science and religion—they are natural antagonists and the history of their interaction records the increasing advance of science into intellectual territory formerly occupied by religion. In this context, what could Burnet represent but a futile finger in a truly crumbling dike?

But the actual relationship between religion and science is far more complex and varied. Often, religion has actively encouraged science. If there is any consistent enemy of science, it is not religion, but irrationalism. Indeed, Burnet, the divine, fell prey to the same forces that persecuted Scopes, the science teacher, almost three centuries later in Tennessee. By examining Burnet's case in a time and a world so different from our own, we may gain a broader understanding of the persistent forces arrayed against science.

I will begin by sketching Burnet's theory. From our point of view, it will appear so silly and contrived that a role for Burnet among dogmatic antiscientists will seem almost inescapable. But I will then examine his methods of inquiry in order to place him among the scientific rationalists of his time. In noting his persecution by dogmatic theology, we merely watch the Huxley-Wilberforce debate or the creation controversy of California played again by the same actors in different garb.

Burnet began his inquiry to determine where the waters of Noah's flood came from. He was convinced that the modern oceans could not drown the earth's mountains. "I can as soon believe," wrote a contemporary, "that a man could be drowned in his own spittle as that the world should be deluged by the water in it." Burnet rejected the idea that Noah's flood might have been a merely local event, falsely extended by witnesses who could not have traveled widely—for that would contravene the authority of sacred scripture. But he rejected even more strongly the notion that God had simply created the extra water as a miracle—for that would dispute the rational world of science. He was led, instead, to the following account of earth history.

From the chaos of the primeval void, our earth precipitated as a perfectly ordered sphere. Its materials sorted themselves according to their densities. Heavy rocks and metals formed a spherical core at the center with a liquid layer above and a sphere of volatiles above the liquid. The volatile layer consisted mostly of air, but it also included terrestrial particles. These precipitated in time to form a perfectly smooth, featureless earth atop the liquid layer.

> In this smooth Earth were the first scenes of the world, and the first generation of Mankind; it had the Beauty of Youth and blooming Nature, fresh and fruitful, and not a Wrinkle, Scar or Fracture in all its body; no Rocks

nor Mountains, no hollow Caves, nor gaping Channels, but even and uniform all over.

There were no seasons amidst this original perfection, for the earth's axis stood bolt upright and the Garden of Eden, conveniently situated in a middle latitude, enjoyed a perpetual spring.

But the earth's own evolution required the destruction of this earthly paradise, and it happened naturally just when disobedient mankind required punishment. Rainfall was light, and the earth began to dry up and crack. The sun's heat vaporized some of the water below the surface. It rose through the cracks, clouds formed, and the rains began. But even forty days and nights could not supply enough water, and more had to rise from the abyss. The falling rain sealed the cracks, forming a pressure cooker without a relief valve as the vaporizing water below pushed upward. The pressure built, and the surface finally burst, causing floods, tidal waves, and the rupture and displacement of the earth's original surface to form mountains and ocean basins. So violent were these disruptions that the earth was wrenched to its current axial tilt (*vide* Velikovsky—essay 19). The waters finally retreated to the abyssal caverns, leaving "a gigantic and hideous ruin . . . a broke and confused heap of bodies." Man, alas, had been made for Eden, and the patriarchal life-span of circa nine-hundred years declined more than tenfold.

And so, according to Reverend Thomas, we inhabitants of a "dirty little planet" await its transformation as promised by scripture and reasoned from planetary physics. The earth's volcanoes will erupt all at once, and the universal conflagration will begin. Protestant Britain, with its reserves of coal (then largely unmined) will burn with a fury, but the fire will surely start in Rome, the papist home of antichrist. The charred particles will precipitate slowly back to earth, forming once again a perfect sphere without relief. And so the one-thousand-year reign of Christ will commence. At its end, the giants Gog and Magog will appear, forcing a new battle between good and evil. The saints will ascend to the bosom of Abraham, and the earth, having run its course, will become a star.

Utterly fantastic? Sure, for 1975; but not for 1681. In fact, for his own time, Burnet was a rationalist, upholding the primacy of Newton's world in an age of faith. For Burnet's primary concern was to render earth history not by miracles or divine caprice, but by natural, physical processes. Burnet's tale may be fanciful, but his actors are the ordinary physical forces of desiccation, evaporation, precipitation, and combustion. To be sure, he believed that the facts of earth history were given unambiguously in scripture, yet they must be consistent with science, lest God's words be opposed to his works. Reason and revelation are two infallible guides to truth, but

> 'tis a dangerous thing to ingage the authority of Scripture in disputes about the Natural World, in opposition to Reason; lest Time, which brings all things to light, should discover that to be evidently false which we had made Scripture to assert.

Moreover, Burnet's God is not the continuous and miraculous actor of a prescientific age, but Newton's imperial clockwinder who, having created matter and ordained its laws, let nature run its own course:

> We think him a better Artist that makes a Clock that strikes regularly at every hour from the Springs and Wheels which he puts in the work, than he that hath so made his Clock that he must put his finger to it every hour to make it strike; And if one should contrive a piece of Clock-work so that it should beat all the hours and make all its motions regularly for such a time, and that time being come, upon a signal given, or a Spring toucht, it should of its own accord fall all to pieces; would not this be look'd upon as a piece of greater Art, than if the Workman came to that time prefixt, and with a great Hammer beat it into pieces?

I do not, of course, argue that Burnet was a scientist in any modern sense of the term. He performed no experiments and he made no observations of rocks and fossils (although several of his contemporaries did). He used a method of "pure" (we would say armchair) reason, and he wrote with as much confidence about an unobservable future as about a verifiable past. Likewise, his procedure is followed by no modern scientist that I know, with the exception of Immanuel Velikovsky (see essay 19)—for Burnet assumed the truth of scripture and fashioned a physical mechanism to make it happen, just as Velikovsky invented a new planetary physics to preserve the literal account of ancient records.

Yet Burnet was no pillar of the theistic establishment. In fact, he got himself into considerable trouble over the sacred theory. In the best tones of the Inquisition, the bishop of Hereford attacked Burnet's reliance on reason: "Either his Brain is crakt with overlove of his own Invention, or his Heart is rotten with some evil design"—that is, the subversion of the church. In a classic statement of antiscience, another clerical critic remarked: "Though we have Moses, yet I believe we must stay [wait] for Elias, to make out to us, the true Philosophical modus of the Creation and Deluge." (The biblical reference is to Elijah, who will return to herald the Messiah's advent—that is, science cannot discuss these questions and we must await some future revelation for their solution.) John Keill, an Oxford mathematician, argued that Burnet's natural explanations were dangerous because they encouraged a belief that God is superfluous.

Nonetheless, Burnet prospered for a time. He became Clerk of the Closet at the court of William III. (That title is not a fancy name for latrine cleaners but a designation for the royal confessor—the closet being a chapel for the king's private devotions.) Rumor has it that he was even considered as a possible successor to the Archbishop of Canterbury. But Burnet finally went too far. In 1692, he published a work advocating an allegorical interpretation of the six days of Genesis—and he promptly lost his job, despite his profuse apologies for any unintended offense.

It was the dogmatists and antirationalists who got Burnet in the end, not the theists (there were no reputable atheists outside the closet in seventeenth-century England). One hundred years later, the same men made Buffon retract his theory of the earth's antiquity. One hundred fifty years after that, they unleashed a pompous, three-time loser against John Scopes. Today, using the liberal rhetoric of equal time, they are trying to drive evolutionary theory from the nation's textbooks.

Science, to be sure, has transgressed as well. We have persecuted dissenters, resorted to catechism, and tried to extend our authority to a moral sphere where it has no force. Yet without a commitment to science and rationality in its proper domain, there can be no solution to the problems that engulf us. Still, the Yahoos never rest.

18 Uniformity and Catastrophe

THE GIDEON SOCIETY—those purveyors of spiritual comfort to a mobile nation—persist in recording the date of creation as 4,004 B.C. in their marginal annotation to Genesis 1. Geologists believe that our planet is at least a million times more ancient—some 4 1/2 billion years old.

Each of the major sciences has contributed an essential ingredient to our long retreat from an initial belief in our own cosmic importance. Astronomy defined our home as a small planet tucked away in one corner of an average galaxy among millions; biology took away our status as paragons created in the image of God; geology gave us the immensity of time and taught us how little of it our own species has occupied.

In 1975, we celebrated the centenary of the death of Charles Lyell, conventional hero of the geologic revolution—"the mirror of all that really mattered in geologic thought," according to one recent biographer. The standard account of Lyell's accomplishment runs in the following way: in the early nineteenth century, geology was dominated by the catastrophists—theological apologists who sought to compress the geologic record into the strictures of biblical chronology. To do this, they imagined a profound discordance between past and present modes of change. The present may run slowly and gradually as waves and rivers do their work; the events of the past were abrupt and cataclysmic

—for how else could they fit into a few thousand years? Mountains were raised in a day, and canyons opened at once. Thus, the Lord interposed his will to break the rule of natural law and place the past outside the sphere of scientific explanation. Loren Eiseley writes: "[Lyell] entered the geological domain when it was a weird, half-lit landscape of gigantic convulsions, floods and supernatural creations and extinctions of life. Distinguished men had lent the power of their names to these theological fantasies."

In 1830, Lyell published the first volume of his revolutionary *Principles of Geology.* He, so the standard story goes, boldly proclaimed that time had no limit. Having removed this fundamental constraint, he advocated a philosophy of "uniformitarianism"—the doctrine that made geology a science. Natural law is invariant. With so much time, we need invoke no more than the slow and steady operation of present causes to produce the entire panorama of past events. The present is the key to the past.

This tale of Lyell's role is no different from most standard accounts in the history of science: it is long on inspiration and rather short on accuracy.

A few months ago, while browsing through the stacks of Harvard's ancient library, I discovered Louis Agassiz's annotated copy of Lyell's *Principles of Geology* (more things are buried in libraries than this world dreams of). Agassiz was America's leading biologist and also her staunchest catastrophist. Yet his marginalia include an impossible contradiction if we accept the standard account of Lyell's achievement. Agassiz's penciled annotations include all the standard critiques of the catastrophist school. They record, in particular, Agassiz's conviction that the summation of present causes over geologic time cannot account for the magnitude of some past events; a notion of cataclysm, he believes, is still required. Nonetheless, he writes as his final assessment: "Mr. Lyell's Principles of Geology is certainly the most important work that has appeared on the whole of this science since it has merited its name." (It occurred to me that Agassiz might have been citing someone else's assessment from a published review; but I have consulted with several historians and we have concluded that his annotation reflects his own opinion.)

If catastrophists wore the black moustaches, if uniformitarians sported silver stars and white hats, and if Lyell was the sharp-shooting sheriff who kicked all the baddies out of town—the Manichaean or western-movie version of the history of science—then Agassiz's statements are absurd, for how could a wrongdoer at liberty praise the sheriff so obsequiously? Either the western script is wrong or Agassiz was crazy.

Why, then, did Agassiz praise Lyell? To answer that question, I must analyze Lyell's so-called uniformitarianism, in order to argue that modern geology is really a blend of concepts from both Lyell and the catastrophists.

Charles Lyell was a lawyer by profession, and his book is one of the most brilliant briefs ever published by an advocate. It is a mélange of precise documentation, incisive argument, and a few of the "quiddities, quillets [quibbles] . . . and tricks" that Hamlet ascribed to the profession when he exhumed a lawyer's skull from the graveyard. Lyell relied upon two bits of cunning to establish his uniformitarian view as the only true geology.

First, he set up a straw man to demolish. By 1830, no serious scientific catastrophist believed that cataclysms had a supernatural cause or that the earth was 6,000 years old. Yet, these notions were held by many laymen, and they were advocated by some quasi-scientific theologians. A scientific geology required their defeat, but they had been routed within the profession by both catastrophists and uniformitarians. Agassiz praised Lyell because he brought a geologic consensus so forcefully to the public.

It is not Lyell's fault that later generations accepted his straw man as an accurate representation of the *scientific* opposition to uniformitarianism. Yet all of the great nineteenth-century catastrophists—Cuvier, Agassiz, Sedgwick, and Murchison in particular—accepted an earth of great antiquity, and they all sought a natural basis for the cataclysmic changes that occurred in the past. A 6,000-year-old earth does require a belief in catastrophes to compress the geologic record into so short a time. But the converse is decidedly not true: a belief in catastrophes does not dictate a 6,000-year-old earth. The earth might be 4.5 or, for that matter, 100 billion years old and still build its mountains with great rapidity.

In fact, the catastrophists were much more empirically minded than Lyell. The geologic record does seem to record catastrophes: rocks are fractured and contorted; whole faunas are wiped out (see essay 16). To circumvent this literal appearance, Lyell imposed his imagination upon the evidence. The geologic record, he argued, is extremely imperfect and we must interpolate into it what we can reasonably infer but cannot see. The catastrophists were the hard-nosed empiricists of their day, not the blinded theological apologists.

Secondly, Lyell's "uniformity" is a hodgepodge of claims. One is a methodological statement that must be accepted by any scientist, catastrophist and uniformitarian alike. Others are substantive notions that have since been tested and abandoned. Lyell gave them a common name and pulled a consummate fast one: he tried to slip the substantive claim by with an argument that the methodological proposition had to be accepted, lest "we see the ancient spirit of speculation revived, and a desire manifested to cut, rather than patiently to untie, the Gordian knot."

Lyell's concept of uniformity has four major, and very different, components:

1. Natural laws are constant (uniform) in space and time. As John Stuart Mill showed, this is not a statement about the world; it is an a priori claim of method that scientists must make in order to proceed with any analysis of the past. If the past is capricious, if God violates natural law at will, then science cannot unravel history. Agassiz and the catastrophists agreed; they, too, sought a natural cause for cataclysms and praised Lyell's basic defense of science against theological intrusion.

2. Processes now operating to mold the earth's surface should be invoked to explain the events of the past (uniformity of process through time). Only present processes can be directly observed. Therefore, we are better off if we can explain past events as a result of processes still acting. This

again is not an argument about the world; it is a statement about scientific procedure. And again, no scientist disagreed. Agassiz and the catastrophists also preferred present processes, and they applauded Lyell's exquisite documentation of how much these processes can accomplish. Their disagreement concerned another issue. Lyell believed that present processes were sufficient to explain the past; catastrophists held that present processes should always be preferred, but that some past events required the inference of causes no longer acting or acting now at markedly slower rates.

3 . Geologic change is slow, gradual, and steady, not cataclysmic or paroxysmal (uniformity of rate). Here we finally encounter a substantive claim that can be tested—and a point of real difference between Agassiz and Lyell. Modern geologists would argue that Lyell's view has largely prevailed, although they would also point out that his original insistence on a near uniformity of rate was stifling to the imagination. (Lyell, for example, never accepted the glacial theory that Agassiz developed; he would not concede that amounts of ice and rates of flow had been so different in the past.)

4. The earth has been fundamentally the same since its formation (uniformity of configuration). This last component of Lyell's uniformity is rarely discussed. After all, it is an empirical claim, and largely an incorrect one at that—and who wants to expose the false steps of a hero? Yet I believe that this uniformity was closest to Lyell's heart and most central to his concept of the earth. Newton's earth revolves endlessly about its star with no direction to its history. One moment is like all moments. Could not such a grand vision apply to the geological record of our planet as well? Land and sea might change their positions, but land and sea exist through time in roughly the same proportion; species come and go, but the mean complexity of life remains forever constant. Endless change in detail, ceaseless constancy in aspect—a dynamic steady state, to use today's jargon of information theory.

Lyell's vision led him to propose, contrary to all evidence, that mammals would be found in the earliest fossiliferous beds. To reconcile the appearance of direction with dynamic constancy in the history of life, he supposed that the entire fossil record represents but one part of a "great year"—a grand cycle that will occur again when "the huge iguanodon might reappear in the woods, and the ichthyosaur in the sea, while the pterodactyle [sic] might flit again through umbrageous groves of tree ferns."

The catastrophists took the literal view. They saw direction in the history of life, and they believed it. In retrospect, they were right.

Most geologists would tell you that their science represents the total triumph of Lyell's uniformity over unscientific catastrophism. Lyell's brief won the victory for his name, but modern geology is really an even mixture of two scientific schools—Lyell's original, rigid uniformitarianism and the scientific catastrophism of Cuvier and Agassiz. We accept Lyell's first two uniformities, but so did the catastrophists. Lyell's third uniformity, appropriately derigidified, is his great substantive contribution; his fourth (and most important) uniformity has been graciously forgotten.

Yet there is much to be said for Lyell's vision of steady state. A dynamic constancy may seem fundamentally at odds with clearly directional aspects of the history of life and the earth. But medieval Christianity could encompass both views in its concept of history. In the stained glass of Chartres, human history is displayed as a linear sequence beginning in the north transept and running around the nave to the south transept—a directional process: one creation, one coming of Christ, one resurrection of the dead. But correspondence also pervades the system, giving a timelessness to apparent direction. The New Testament is a replay of the Old. Mary is like the burning bush because both held within themselves the fire of God, yet were not consumed. Christ is like Jonah because both arose again after three days *in extremis.* The two visions—directionalism and dynamic constancy—are not irreconcilable. Geology, too, might seek their creative synthesis.

19 Velikovsky in Collision

NOT LONG AGO, Venus emerged from Jupiter, like Athena from the brow of Zeus—literally! It then assumed the form and orbit of a comet. In 1500 B.C., at the time of the Jewish exodus from Egypt, the earth passed twice through Venus's tail, bringing both blessings and chaos; manna from heaven (or rather from hydrocarbons of a cometary tail) and the bloody rivers of the Mosaic plagues (iron from the same tail). Continuing its erratic course, Venus collided with (or nearly brushed) Mars, lost its tail, and hurtled to its present orbit. Mars then left its regular position and almost collided with the earth in about 700 B.C. So great were the terrors of these times, and so ardent our collective desire to forget them, that they have been erased from our conscious minds. Yet they lurk in our inherited and unconscious memory, causing fear, neurosis, aggression, and their social manifestations as war.

This may sound like the script of a very poor, late-late movie on TV; nonetheless, it represents the serious theory of Immanuel Velikovsky's *Worlds in Collision.* And Velikovsky is neither crank nor charlatan—although to state my opinion and to quote one of my colleagues, he is at least gloriously wrong.

Worlds in Collision, published twenty-five years ago, continues to engender intense debate. It also has spawned a series of issues peripheral to the purely scientific arguments. Velikovsky was surely ill treated by certain academics who sought to suppress the publication of his work. But a man does not attain the status of Galileo merely because he is persecuted; he must also be right. The scientific and sociological issues are separate. And then, times and the treatment of heretics have changed. Bruno was burned to death; Galileo, after viewing the instruments of torture, languished under house arrest. Velikovsky won both publicity and royalties. Torquemada was evil; Velikovsky's academic enemies, merely foolish.

As startling as his specific claims may be, I am more interested in Velikovsky's unorthodox method of inquiry and physical theory. He begins with the working hypothesis that all stories reported as direct observation in the ancient chronicles are strictly true—if the Bible reports that the sun stood still, then it did (as the tug of Venus briefly halted the earth's rotation). He then attempts to find some physical explanation, however bizarre, that would render all these stories both mutually consistent and true. Most scientists would do exactly the opposite in using the limits of physical possibility to judge which of the ancient legends *might* be literally accurate. (I devoted essay 17 to the last important scientific work that used Velikovsky's method—Thomas Burnet's *Sacred Theory of the Earth,* first published in the 1680s.) Secondly, Velikovsky is well aware that the laws of Newton's universe, where forces of gravitation rule the motion of large objects, will not allow planets to wander. Thus, he proposes a fundamentally new physics of electromagnetic forces for large bodies. In short, Velikovsky would rebuild the science of celestial mechanics to save the literal accuracy of ancient legends.

Having devised a cataclysmic theory of human history, Velikovsky then sought to generalize his physics by extending it throughout geologic time. In 1955 he published *Earth in Upheaval,* his geologic treatise. With Newton and modern physics already under siege, he now took on Charles Lyell and modern geology. He reasoned that if wandering planets had visited us twice within a mere 3,500 years, then the history of the earth should be marked by its catastrophes, not by the slow and gradual change that Lyell's uniformitarianism requires.

Velikovsky scoured the geologic literature of the past hundred years for records of cataclysmic events—floods, earthquakes, volcanoes, mountain building, mass extinctions, and shifts of climate. Finding these aplenty, he sought a common cause:

> Sudden the agent must have been and violent; recurrent it must have been, but at highly erratic intervals; and it must have been of titanic power.

Not surprisingly, he invoked the electromagnetic forces of celestial bodies external to the earth. In particular, he argues that these forces rapidly perturb the earth's rotation—literally turning the earth over in extreme cases and exchanging poles for equators. Velikovsky offers a rather colorful account of the effects that might accompany such a sudden shift in the earth's axis of rotation:

> At that moment an earthquake would make the globe shudder. Air and water would continue to move through inertia; hurricanes would sweep the earth and the seas would rush over continents. . . . Heat would be developed, rocks would melt, volcanoes would erupt, lava would flow from fissures in the ruptured ground and cover vast areas. Mountains would spring up from the plains.

If the testimony of human narrators provided the evidence for *Worlds in Collision,* then the geologic record itself must suffice for *Earth in Upheaval.* Velikovsky's entire argument hinges on his reading of geological literature. This, I feel, he does rather badly and carelessly. I will focus upon the general faults of his procedure, not the refutation of specific claims.

First, the assumption that similarity of form reflects simultaneity of occurrence: Velikovsky discusses the fossil fishes of the Old Red Sandstone, a Devonian formation in England (350–400 million years old). He cites evidence of violent death—contortion of the body, lack of predation, even signs of "surprise and terror" engraved forever on fossil faces. He infers that some sudden catastrophe must have extirpated all these fishes; yet, however unpleasant the death of any individual, these fishes are distributed through hundreds of feet of sediments that record several million years of deposition! Likewise, the craters of the moon are similar in appearance, and each one formed by the sudden impact of a meteorite. Yet this influx spans billions of years, and Velikovsky's favored hypothesis of simultaneous origin by bubbling on the surface of a molten moon has been conclusively disproved by the Apollo landings.

Second, the assumption that events are sudden because their effects are large: Velikovsky writes graphically about the hundreds of feet of ocean water that were evaporated to form the great Pleistocene ice sheets. He can envisage the process only as a result of oceanic boiling followed by a general refrigeration:

> An unusual sequence of events was necessary: the oceans must have steamed and the vaporized water must have fallen as snow in latitudes of temperate climates. This sequence of heat and cold must have taken place in quick succession.

Yet glaciers are not built overnight. They formed "rapidly" by geological standards, but the few thousand years of their growth allowed ample time for the gradual accumulation of snow by new precipitation supplied each year. One need not make the oceans steam; it still snows in northern Canada.

Third, the inference of worldwide events from local catastrophes: no geologist has ever denied that *local* catastrophes occur by flooding, earthquake, or volcanic eruption. But these events have nothing to do, one way or the other, with Velikovsky's notion of global catastrophe caused by sudden shifts in the earth's axis. Nevertheless, most of Velikovsky's "examples" are just such local events combined with an unwarranted extrapolation to global impact. He writes, for example, of the Agate Springs Quarry of Nebraska—a local mammalian "graveyard" containing the bones (according to one estimate) of nearly 20,000 large animals. But, this large aggregation may not record a catastrophic event at all—rivers and oceans can gradually accumulate vast quantities of bone and shell (I have walked on beaches composed entirely of large shells and coral rubble). Also, even if a local flood drowned these animals, we have no evidence that their contemporary brethren on other continents were the least bit bothered.

Fourth, the exclusive use of outdated sources: before 1850, most geologists invoked general catastrophes as the major agent of geologic change. These men were not stupid, and they argued their position with some cogency. If we read only their works, their conclusions seem to follow. Velikovsky's entire discussion of the catastrophic death of European fossil fishes cites only the works of Hugh Miller in 1841 and of William Buckland in 1820 and 1837. Surely the past hundred years, with its voluminous literature, contains something worth noting. Likewise, Velikovsky relies on John Tyndall's work of 1883 for his meteorological notions about the origin of ice ages. Yet scarcely any subject has been more actively discussed in geological circles during this century.

Fifth, carelessness, inaccuracy, and sleight of hand: *Earth in Upheaval* is studded with minor errors and half-truths, unimportant in themselves, but reflecting either a cavalier attitude toward the geologic literature or, more simply, a failure to understand it. Thus, Velikovsky attacks the uniformitarian postulate that present causes can explain the past by arguing that no fossils are forming today. Anyone who has dug old bones from lake beds or shells from beaches knows that this claim is simply absurd. Likewise, Velikovsky refutes Darwinian gradualism with an argument "that some organisms, like foraminifera, survived all geological ages without participating in evolution." This claim was occasionally made in older literature written before anyone had seriously studied these single-celled creatures. But no one has maintained it since J. A. Cushman's voluminous descriptive work of the 1920s. Finally, we learn that igneous rocks—granite and basalt—"have embedded in them numberless living organisms." This is news to me and to the entire profession of paleontology.

But all these criticisms pale to insignificance before the most conclusive refutation of Velikovsky's examples—their explanation as consequences of continental drift and plate tectonics. And here Velikovsky is not to blame at all. He has merely fallen victim—as have so many others with the most orthodox among previously cherished opinions—to this great revolution in geological thought. In *Earth in Upheaval,* Velikovsky quite reasonably rejected continental drift as an alternate explanation for the most important phenomena supporting his catastrophic theory. And he rejected it for the reason then most commonly cited among geologists—the lack of a mechanism to move the continents. That mechanism has now been provided with the verification of sea-floor spreading (see essays 16 and 20). The African rift is not a crack formed when the earth turned over rapidly; it is a part of the earth's rift system, a junction between two crustal plates. The Himalayas did not rise when the earth shifted but when the Indian plate slowly pushed into Asia. The Pacific volcanoes, a "ring of fire," are not the product of melting during the last axial displacement; they mark the boundary between two plates. There are fossil corals in polar regions, coal in Antarctica, and evidence of Permian glaciation in tropical South America. But the earth need not turn over to explain all this; the continents have only to drift from different climatic realms into their present positions.

Ironically, Velikovsky has lost more to plate tectonics than his mechanism of axial shifting; he has probably lost the entire rationale for his catastrophist position. As Walter Sullivan argues in his recent book on continental drift, the theory of plate tectonics has provided a stunning confirmation of uniformitarian preferences for ascribing past events to present causes acting without great deviation from their current intensity. For the plates are actively moving today, carrying their continents with them. And the sweeping panorama of attendant events—the worldwide belt of earthquakes and volcanoes, the collision of continents, the mass extinctions of faunas (see essay 16)—can be explained by the continuous movement of these giant plates at rates of only a few centimeters a year.

The Velikovsky affair raises what is perhaps the most disturbing question about the public impact of science. How is a layman to judge rival claims of supposed experts? Any person with a gift for words can spin a persuasive argument about any subject not in the domain of a reader's personal expertise. Even von Daniken sounds good if you just read *Chariots of the Gods.* I am in no position to judge the historical argument of *Worlds in Collision.* I know little of celestial mechanics and even less about the history of the Egyptian Middle Kingdom (although I have heard experts howl about Velikovsky's unorthodox chronology). I do not wish to assume that the nonprofessional must be wrong. Yet when I see how poorly Velikovsky uses the data I am familiar with, then I must entertain doubts about his handling of material unfamiliar to me. But what is a person who knows neither astronomy, Egyptology, nor geology to do—especially when faced with a hypothesis so intrinsically exciting and a tendency, shared, I suspect, by all of us, to root for the underdog?

We know that many fundamental beliefs of modern science arose as heretical speculations advanced by nonprofessionals. Yet history provides a biased filter for our judgment. We sing praises to the unorthodox hero, but for each successful heretic, there are a hundred forgotten men who challenged prevailing notions and lost. Who among you has ever heard of Eimer, Cuénot, Trueman, or Lang—the primary supporters of orthogenesis (directed evolution) against the Darwinian tide? Still, I will continue to root for heresy preached by the nonprofessional. Unfortunately, I don't think that Velikovsky will be among the victors in this hardest of all games to win.

20 The Validation of Continental Drift

AS THE NEW Darwinian orthodoxy swept through Europe, its most brilliant opponent, the aging embryologist Karl Ernst von Baer, remarked with bitter irony that every triumphant theory passes through three stages: first it is dismissed as untrue; then it is rejected as contrary to religion; finally, it is accepted as dogma and each scientist claims that he had long appreciated its truth.

I first met the theory of continental drift when it labored under the inquisition of stage two. Kenneth Caster, the only major American paleontologist who dared to support it openly, came to lecture at my alma mater, Antioch College. We were scarcely known as a bastion of entrenched conservatism, but most of us dismissed his thoughts as just this side of sane. (Since I am now in von Baer's third stage, I have the distinct memory that Caster sowed substantial seeds of doubt in my own mind.) A few years later, as a graduate student at Columbia University, I remember the a priori derision of my distinguished stratigraphy professor toward a visiting Australian drifter. He nearly orchestrated the chorus of Bronx cheers from a sycophantic crowd of loyal students. (Again, from my vantage point in the third stage, I recall this episode as amusing, but distasteful.) As a tribute to my professor, I must record that he experienced a rapid conversion just two years later and spent his remaining years joyously redoing his life's work.

Today, just ten years later, my own students would dismiss

with even more derision anyone who denied the evident truth of continental drift—a prophetic madman is at least amusing; a superannuated fuddy-duddy is merely pitiful. Why has such a profound change occurred in the short space of a decade?

Most scientists maintain—or at least argue for public consumption—that their profession marches toward truth by accumulating more and more data, under the guidance of an infallible procedure called "the scientific method." If this were true, my question would have an easy answer. The facts, as known ten years ago, spoke against continental drift; since then, we have learned more and revised our opinions accordingly. I will argue, however, that this scenario is both inapplicable in general and utterly inaccurate in this case.

During the period of nearly universal rejection, direct evidence for continental drift—that is, the data gathered from rocks exposed on our continents—was every bit as good as it is today. It was dismissed because no one had devised a physical mechanism that would permit continents to plow through an apparently solid oceanic floor. In the absence of a plausible mechanism, the idea of continental drift was rejected as absurd. The data that seemed to support it could always be explained away. If these explanations sounded contrived or forced, they were not half so improbable as the alternative—accepting continental drift. During the past ten years, we have collected a new set of data, this time from the ocean basins. With these data, a heavy dose of creative imagination, and a better understanding of the earth's interior, we have fashioned a new theory of planetary dynamics. Under this theory of plate tectonics, continental drift is an inescapable consequence. The old data from continental rocks, once soundly rejected, have been exhumed and exalted as conclusive proof of drift. In short, we now accept continental drift because it is the expectation of a new orthodoxy.

I regard this tale as typical of scientific progress. New facts, collected in old ways under the guidance of old theories, rarely lead to any substantial revision of thought. Facts do not "speak for themselves"; they are read in the light of theory. Creative thought, in science as much as in the arts, is the motor of changing opinion. Science is a quintessentially human activity, not a mechanized, robotlike accumulation of objective information, leading by laws of logic to inescapable interpretation. I will try to illustrate this thesis with two examples drawn from the "classical" data for continental drift. Both are old tales that had to be undermined while drift remained unpopular.

I. The late Paleozoic glaciation. About 240 million years ago, glaciers covered parts of what is now South America, Antarctica, India, Africa, and Australia. If continents are stable, this distribution presents some apparently insuperable difficulties:

> A. The orientation of striae in eastern South America indicates that glaciers moved onto the continent from what is now the Atlantic Ocean (striae are scratches on bedrock made by rocks frozen into glacier bottoms as they pass over a surface). The world's oceans form a single system, and transport of heat from tropical areas guarantees that no major part of the open ocean can freeze.
>
> B. African glaciers covered what are now tropical areas.
>
> C. Indian glaciers must have grown in semitropical regions of the Northern hemisphere; moreover, their striae indicate a source in tropical waters of the Indian Ocean.
>
> D. There were no glaciers on any of the northern continents. If the earth got cold enough to freeze tropical Africa, why were there no glaciers in northern Canada or Siberia?

All these difficulties evaporate if the southern continents (including India) were joined together during this glacial period, and located farther south, covering the South Pole; the South American glaciers moved from Africa, not an open ocean; "tropical" Africa and "semitropical" India were near the South Pole; the North Pole lay in the middle of a major ocean, and glaciers could not develop in the Northern Hemisphere. Sounds good for drift; indeed, no one doubts it today.

II. The distribution of Cambrian trilobites (fossil arthropods living 500 to 600 million years ago). The Cambrian trilobites of Europe and North America divided themselves into two rather different faunas with the following peculiar distribution on modern maps. "Atlantic" province trilobites lived all over Europe and in a few very local areas on the far eastern border of North America—eastern (but not western) Newfoundland and southeastern Massachusetts, for example. "Pacific" province trilobites lived all over America and in a few local areas on the extreme western coast of Europe —northern Scotland and northwestern Norway, for example. It is devilishly difficult to make any sense of this distribution if the two continents always stood 3,000 miles apart.

But continental drift suggests a striking resolution. In Cambrian times, Europe and North America were separated: Atlantic trilobites lived in waters around Europe; Pacific trilobites in waters around America. The continents (now including sediments with entombed trilobites) then drifted toward each other and finally joined together. Later, they split again, but not precisely along the line of their previous junction. Scattered bits of ancient Europe, carrying Atlantic trilobites, remained at the easternmost border of North America, while a few pieces of old North America stuck to the westernmost edge of Europe.

Both examples are widely cited as "proofs" of drift today, but they were soundly rejected in previous years, not because their data were any less complete but only because no one had devised an adequate mechanism to move continents. All the original drifters imagined that continents plow their way through a static ocean floor. Alfred Wegener, the father of continental drift, argued early in our century that gravity alone could put continents in motion. Continents drift slowly westward, for example, because attractive forces of the sun and moon hold them up as the earth rotates underneath them. Physicists responded with derision and showed mathematically that gravitational forces are far too weak to power such a monumental peregrination. So Alexis du Toit, Wegener's South African champion, tried a different tack. He argued for a local, radioactive melting of oceanic floor at continental borders, permitting the continents to glide through. This *ad hoc* hypothesis added no increment of plausibility to Wegener's speculation.

Since drift seemed absurd in the absence of a mechanism,

orthodox geologists set out to render the impressive evidence for it as a series of unconnected coincidences.

In 1932, the famous American geologist Bailey Willis strove to make the evidence of glaciation compatible with static continents. He invoked the deus ex machina of "isthmian links"—narrow land bridges flung with daring abandon across 3,000 miles of ocean. He placed one between eastern Brazil and western Africa, another from Africa all the way to India via the Malagasy Republic, and a third from Vietnam through Borneo and New Guinea to Australia. His colleague, Yale professor Charles Schuchert, added one from Australia to Antarctica and another from Antarctica to South America, thus completing the isolation of a southern ocean from the rest of the world's waters. Such an isolated ocean might freeze along its southern margin, permitting glaciers to flow across into eastern South America. Its cold waters would also nourish the glaciers of southern Africa. The Indian glaciers, located above the equator 3,000 miles north of any southern ice, demanded a separate explanation. Willis wrote: "No direct connection between the occurrences can reasonably be assumed. The case must be considered on the basis of a general cause and the local geographic and topographic conditions." Willis's inventive mind was equal to the task: he simply postulated a topography so elevated that warm, wet southern waters precipitated their product as snow. For the absence of ice in temperate and arctic zones of the Northern Hemisphere, Willis reconstructed a system of ocean currents that permitted him to postulate "a warm, subsurface current flowing northward beneath cooler surface waters and rising in the Artctic as a warm-water heating system." Schuchert was delighted with the resolution provided by isthmian links:

> Grant the biogeographer Holarctis, a land bridge from northern Africa to Brazil, another from South America to Antarctis (it almost exists today), still another from this polar land to Australia and from the latter across the Arafura Sea to Borneo and Sumatra and so on to Asia, plus the accepted means of dispersal along shelf seas and by wind and water currents and migratory birds, and he has all the possibilities needed to explain the life disper-

sion and the land and ocean realms throughout geological time on the basis of the present arrangement of the continents.

The only common property shared by all these land bridges was their utterly hypothetical status; not an iota of direct evidence supported any one of them. Yet, lest the saga of isthmian links be read as a warped fairy tale invented by dogmatists to support an untenable orthodoxy, I point out that to Willis, Schuchert, and any right-thinking geologist of the 1930s, one thing legitimately seemed ten times as absurd as imaginary land bridges thousands of miles long—continental drift itself.

In the light of such highly fertile imaginations, the Cambrian trilobites could present no insuperable problem. The Atlantic and Pacific provinces were interpreted as different environments, rather than different places—shallow water for the Pacific, deeper for the Atlantic. With a freedom to invent nearly any hypothetical geometry for Cambrian ocean basins, geologists drew their maps and hewed to their orthodoxy.

When continental drift came into fashion during the late 1960s, the classical data from continental rocks played no role at all: drift rode in on the coattails of a new theory, supported by new types of evidence. The physical absurdities of Wegener's theory rested on his conviction that continents cut their way through the ocean floor. But how else could drift occur? The ocean floor, the crust of the earth, must be stable. After all, where could it go, if it moved in pieces, without leaving gaping holes in the earth? Nothing could be clearer. Or could it?

"Impossible" is usually defined by our theories, not given by nature. Revolutionary theories trade in the unexpected. If continents must plow through oceans, then drift will not occur; suppose, however, that continents are frozen into the oceanic crust and move passively as pieces of crust shift about. But we just stated that the crust cannot move without leaving holes. Here, we reach an impasse that must be bridged by creative imagination, not just by another field season in the folded Appalachians—we must model the earth in a fundamentally different way.

We can avoid the problem of holes with a daring postulate that seems to be valid. If two pieces of ocean floor move away from each other, they will leave no hole if material rises from the earth's interior to fill the gap. We can go further by reversing the causal implications of this statement: the rise of new material from the earth's interior may be the driving force that moves old sea floor away. But since the earth is not expanding, we must also have regions where old sea floor founders into the earth's interior, thus preserving a balance between creation and destruction.

Indeed, the earth's surface seems to be broken into fewer than ten major "plates," bounded on all sides by narrow zones of creation (oceanic ridges) and destruction (trenches). Continents are frozen into these plates, moving with them as the sea floor spreads away from zones of creation at oceanic ridges. Continental drift is no longer a proud theory in its own right; it has become a passive consequence of our new orthodoxy—plate tectonics.

We now have a new, mobilist orthodoxy, as definite and uncompromising as the staticism it replaced. In its light, the classical data for drift have been exhumed and proclaimed as proof positive. Yet these data played no role in validating the notion of wandering continents; drift triumphed only when it became the necessary consequence of a new theory.

The new orthodoxy colors our vision of all data; there are no "pure facts" in our complex world. About five years ago, paleontologists found on Antarctica a fossil reptile named *Lystrosaurus.* It also lived in South Africa, and probably in South America as well (rocks of the appropriate age have not been found in South America). If anyone had floated such an argument for drift in the presence of Willis and Schuchert, he would have been howled down—and quite correctly. For Antarctica and South America are almost joined today by a string of islands, and they were certainly connected by a land bridge at various times in the past (a minor lowering of sea level would produce such a land bridge today). *Lystrosaurus* may well have walked in comfort, on a rather short journey at that. Yet the *New York Times* wrote an editorial proclaiming, on this basis alone, that continental drift had been proved.

Many readers may be disturbed by my argument for the primacy of theory. Does it not lead to dogmatism and disrespect for fact? It can, of course, but it need not. The lesson of history holds that theories are overthrown by rival theories, not that orthodoxies are unshakable. In the meantime, I am not distressed by the crusading zeal of plate tectonics, for two reasons. My intuition, culturally bound to be sure, tells me that it is basically true. My guts tell me that it's damned exciting—more than enough to show that conventional science can be twice as interesting as anything invented by all the von Dänikens and in all the Bermuda triangles of this and previous ages of human gullibility.

6 Size and Shape, from Churches to Brains to Planets

21 | Size and Shape

Who could believe an ant in theory?
A giraffe in blueprint?
Ten thousand doctors of what's possible
Could reason half the jungle out of being.

JOHN CIARDI'S lines reflect a belief that the exuberant diversity of life will forever frustrate our arrogant claims to omniscience. Yet, however much we celebrate diversity and revel in the peculiarities of animals, we must also acknowledge a striking "lawfulness" in the basic design of organisms. This regularity is most strongly evident in the correlation of size and shape.

Animals are physical objects. They are shaped to their advantage by natural selection. Consequently, they must assume forms best adapted to their size. The relative strength of many fundamental forces (gravity, for example) varies with size in a regular way, and animals respond by systematically altering their shapes.

The geometry of space itself is the major reason for correlations between size and shape. *Simply by growing larger,* any object will suffer continual decrease in relative surface area when its shape remains unchanged. This decrease occurs because volume increases as the cube of length (length × length × length), while surface increases only as the square (length × length): in other words, volume grows more rapidly than surface.

Why is this important to animals? Many functions that depend upon surfaces must serve the entire volume of the body. Digested food passes to the body through surfaces; oxygen is absorbed through surfaces in respiration; the strength of a leg bone depends upon the area of its cross section, but the legs must hold up a body increasing in weight by the cube of its length. Galileo first recognized this principle in his *Discorsi* of 1638, the masterpiece he wrote while

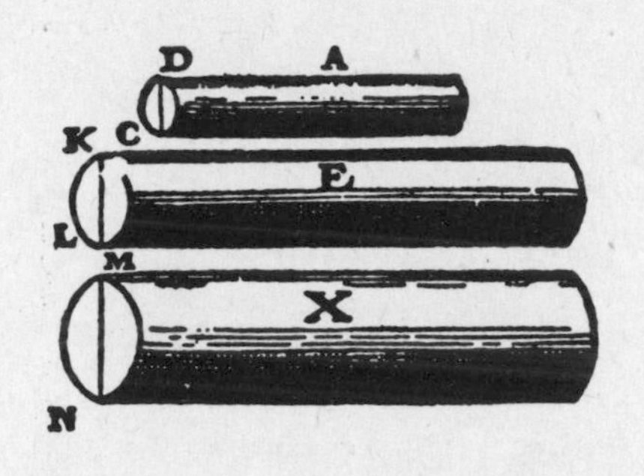

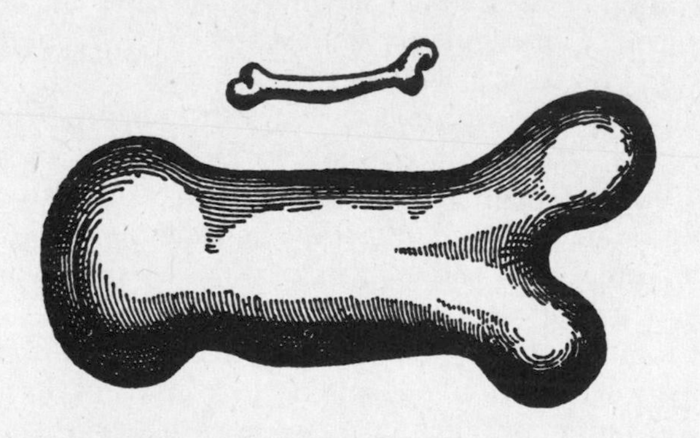

Galileo's original illustration of the relationship between size and shape. To maintain the same strength, large cylinders must be relatively thicker than small ones. For exactly the same reason, large animals have relatively thick leg bones.

under house arrest by the Inquisition. He argued that the bone of a large animal must thicken disproportionately to provide the same relative strength as the slender bone of a small creature.

One solution to decreasing surface has been particularly important in the progressive evolution of large and complex organisms: the development of internal organs. The lung is, essentially, a richly convoluted bag of surface area for the exchange of gases; the circulatory system distributes material to an internal space that cannot be reached by direct diffusion from the external surface of large organisms; the villi of our small intestine increase the surface area available for absorption of food (small mammals neither have nor need them).

Some simpler animals have never evolved internal organs; if they become large, they must alter their entire shape in ways so drastic that plasticity for further evolutionary change is sacrificed to extreme specialization. Thus, a tapeworm may be 20 feet long, but its thickness cannot exceed a fraction of an inch because food and oxygen must penetrate directly from the external surface to all parts of the body.

Other animals are constrained to remain small. Insects breathe through invaginations of their external surface. Oxygen must pass through these surfaces to reach the entire volume of the body. Since these invaginations must be more numerous and convoluted in larger bodies, they impose a limit upon insect size: at the size of even a small mammal, an insect would be "all invagination" and have no room for internal parts.

We are prisoners of the perceptions of our size, and rarely recognize how different the world must appear to small animals. Since our relative surface area is so small at our large size, we are ruled by gravitational forces acting upon our weight. But gravity is negligible to very small animals with high surface to volume ratios; they live in a world dominated by surface forces and judge the pleasures and dangers of their surroundings in ways foreign to our experience.

An insect performs no miracle in walking up a wall or upon the surface of a pond; the small gravitational force pulling it down or under is easily counteracted by surface adhesion.

Throw an insect off the roof and it floats gently down as frictional forces acting upon its surface overcome the weak influence of gravity.

The relative weakness of gravitational forces also permits a mode of growth that large animals could not maintain. Insects have an external skeleton and can only grow by discarding it and secreting a new one to accommodate the enlarged body. For a period between shedding and regrowth, the body must remain soft. A large mammal without any supporting structures would collapse to a formless mass under the influence of gravitational forces; a small insect can maintain its cohesion (related lobsters and crabs can grow much larger because they pass their "soft" stage in the nearly weightless buoyancy of water). We have here another reason for the small size of insects.

The creators of horror and science-fiction movies seem to have no inkling of the relationship between size and shape. These "expanders of the possible" cannot break free from the prejudices of their perceptions. The small people of *Dr. Cyclops, The Bride of Frankenstein, The Incredible Shrinking Man,* and *Fantastic Voyage* behave just like their counterparts of normal dimensions. They fall off cliffs or down stairs with resounding thuds; they wield weapons and swim with olympic agility. The large insects of films too numerous to name continue to walk up walls or fly even at dinosaurian dimensions. When the kindly entomologist of *Them* discovered that the giant queen ants had left for their nuptial flight, he quickly calculated this simple ratio: a normal ant is a fraction of an inch long and can fly hundreds of feet; these ants are many feet long and must be able to fly as much as 1,000 miles. Why, they could be as far away as Los Angeles! (Where, indeed, they were, lurking in the sewers.) But the ability to fly depends upon the surface area of wings, while the weight that must be borne aloft increases as the cube of length. We may be sure that even if the giant ants had somehow circumvented the problems of breathing and growth by molting, their sheer bulk would have grounded them permanently.

Other essential features of organisms change even more

rapidly with increasing size than the ratio of surface to volume. Kinetic energy, in some situations, increases as length raised to the fifth power. If a child half your height falls down, its head will hit with not half, but only 1/32 the energy of yours in a similar fall. A child is protected more by its size than by a "soft" head. In return, we are protected from the physical force of its tantrums, for the child can strike with, not half, but only 1/32 of the energy we can muster. I have long had a special sympathy for the poor dwarfs who suffer under the whip of cruel Alberich in Wagner's *Das Rheingold.* At their diminutive size, they haven't a chance of extracting, with mining picks, the precious minerals that Alberich demands, despite the industrious and incessant leitmotif of their futile attempt.[4]

This simple principle of differential scaling with increasing size may well be the most important determinant of organic shape. J. B. S. Haldane once wrote that "comparative anatomy is largely the story of the struggle to increase surface in proportion to volume." Yet its generality extends beyond life, for the geometry of space constrains ships, buildings, and machines, as well as animals.

Medieval churches present a good testing ground for the effects of size and shape, for they were built in an enormous range of sizes before the invention of steel girders, internal lighting, and air conditioning permitted modern architects to challenge the laws of size. The small, twelfth-century parish church of Little Tey, Essex, England, is a broad, simple rectangular building with a semicircular apse. Light reaches the interior through windows in the outer walls. If we were to build a cathedral simply by enlarging this design, then the area of outer walls and windows would increase as length squared, while the volume that light must reach would increase as length cubed. In other words, the area of the windows would increase far more slowly than the volume that

---

4 | A friend has since pointed out that Alberich, a rather small man himself, would only wield the whip with a fraction of the force we could exert—so things might not have been quite so bad for his underlings.

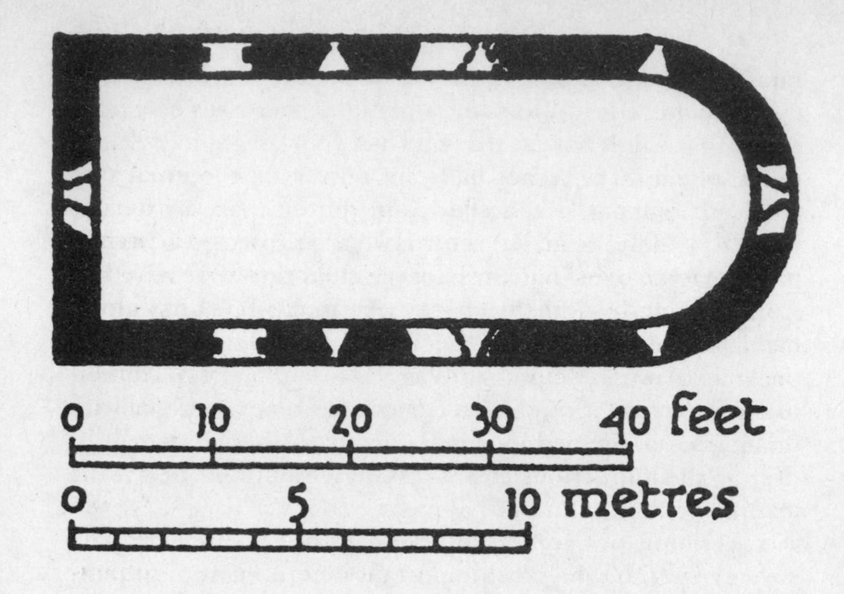

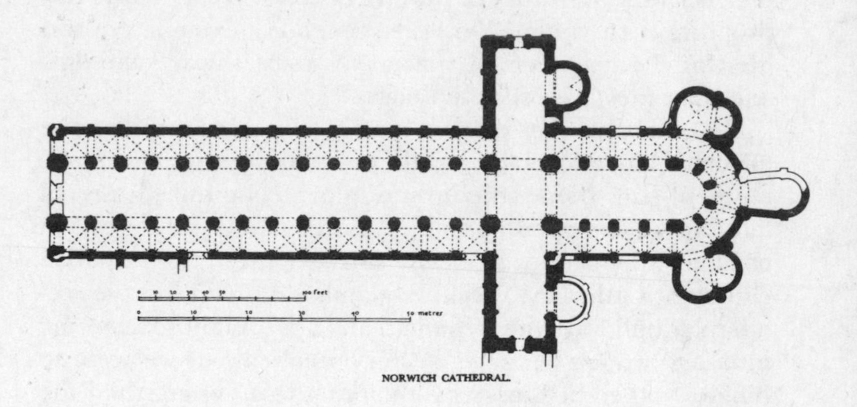

The great range of designs among medieval churches can be attributed partly to size. The twelfth-century parish church of Little Tey, Essex, England, was only 57 feet long and had a simple floor plan, top, while the floor plan for Norwich Cathedral, also twelfth century, shows adaptations—transept, chapels—required for the 450-foot-long building. The need for light and support dictated complex cathedral layouts. (A. W. Clapham, *English Romanesque Architecture: After the Conquest,* Clarendon Press Oxford, 1934. Reprinted with the permission of Oxford University Press)

requires illumination. Candles have limitations; the inside of such a cathedral would have been darker than the deed of Judas. Medieval churches, like tapeworms, lack internal systems and must alter their shape to produce more external surface as they are made larger. In addition, large churches had to be relatively narrow because ceilings were vaulted in stone and large widths could not be spanned without intermediate supports. The chapter house at Batalha, Portugal—one of the widest stone vaults in medieval architecture—collapsed twice during construction and was finally built by prisoners condemned to death.

Consider the large cathedral of Norwich, as it appeared in the twelfth century. In comparison with Little Tey, the rectangle of the nave has become much narrower; chapels have been added to the apse, and a transept runs perpendicular to the main axis. All these "adaptations" increase the ratio of external wall and window to internal volume. It is often stated that transepts were added to produce the form of a Latin cross. Theological motives may have dictated the position of such "outpouchings," but the laws of size required their presence. Very few small churches have transepts. Medieval architects had their rules of thumb, but they had, so far as we know, no explicit knowledge of the laws of size.

Large organisms, like large churches, have very few options open to them. Above a certain size, large terrestrial animals look basically alike—they have thick legs and relatively short, stout bodies. Large medieval churches are relatively long and have abundant outpouchings. The "invention" of internal organs allowed animals to retain the highly successful shape of a simple exterior enclosing a large internal volume; the invention of internal lighting and structural steel has permitted modern architects to design large buildings of essentially cubic form. The limits are expanded, but the laws still operate. No large Gothic church is wider than long; no large animal has a sagging middle like a dachshund.

I once overheard a children's conversation in a New York playground. Two young girls were discussing the size of dogs. One asked: "Can a dog be as large as an elephant?" Her friend responded: "No if it were as big as an elephant, it would look like an elephant." How truly she spoke.

22 | Sizing Up Human Intelligence

A | HUMAN BODIES

"Size," Julian Huxley once remarked, "has a fascination of its own." We stock our zoos with elephants, hippopotamuses, giraffes, and gorillas; who among you was not rooting for King Kong in his various battles atop tall buildings? This focus on the few creatures larger than ourselves has distorted our conception of our own size. Most people think that *Homo sapiens* is a creature of only modest dimensions. In fact, humans are among the largest animals on earth; more than 99 percent of animal species are smaller than we are. Of 190 species in our own order of primate mammals, only the gorilla regularly exceeds us in size.

In our self-appointed role as planetary ruler, we have taken great interest in cataloging the features that permitted us to attain this lofty estate. Our brain, upright posture, development of speech, and group hunting (to name just a few) are often cited, but I have been struck by how rarely our large size has been recognized as a controlling factor of our evolutionary progress.

Despite its low reputation in certain circles, self-conscious intelligence is surely the *sine qua non* of our current status. Could we have evolved it at much smaller body sizes? One day, at the New York World's Fair in 1964, I entered the Hall of Free Enterprise to escape the rain. Inside, prominently displayed, was an ant colony bearing the sign: "Twenty million years of evolutionary stagnation. Why? Because the ant colony is a socialist, totalitarian system." The statement scarcely requires serious attention; nonetheless, I should point out that ants are doing very well for themselves, and that it is their size rather than their social structure that precludes high mental capacity.

In this age of the transistor, we can put radios in watch-cases and bug telephones with minute electronic packages. Such miniaturization might lead us to the false belief that absolute size is irrelevant to the operation of complex machinery. But nature does not miniaturize neurons (or other cells for that matter). The range of cell size among organisms is incomparably smaller than the range in body size. Small animals simply have far fewer cells than large animals. The human brain contains several billion neurons; an ant is constrained by its small size to have many hundreds of times fewer neurons.

There is, to be sure, no established relationship between brain size and intelligence among humans (the tale of Anatole France with a brain of less than 1,000 cubic centimeters vs. Oliver Cromwell with well above 2,000 is often cited). But this observation cannot be extended to differences between species and certainly not to ranges of sizes separating ants and humans. An efficient computer needs billions of circuits and an ant simply cannot contain enough of them because the relative constancy of cell size requires that small brains contain few neurons. Thus, our large body size served as a prerequisite for self-conscious intelligence.

We can make a stronger argument and claim that humans have to be just about the size they are in order to function as they do. In an amusing and provocative article (*American Scientist,* 1968), F. W. Went explored the impossibility of human life, as we know it, at ant dimensions (assuming for the moment that we could circumvent—which we cannot—the problem of intelligence and small brain size). Since weight increases so much faster than surface area as an object gets larger, small animals have very high ratios of surface to volume: they live in a world dominated by surface forces that affect us scarcely at all (see previous essay).

An ant-sized man might don some clothing, but forces of surface adhesion would preclude its removal. The lower limit of drop size would make showering impossible; each drop would hit with the force of a large boulder. If our homunculus managed to get wet and tried to dry off with a towel, he would be stuck to it for life. He could pour no liquid, light no fire (since a stable flame must be several millimeters in length). He might pound gold leaf thin enough to construct a book for his size, but surface adhesion would prevent the turning of pages.

Our skills and behavior are finely attuned to our size. We could not be twice as tall as we are, for the kinetic energy of a fall would then be 16 to 32 times as great, and our sheer weight (increased eightfold) would be more than our legs could support. Human giants of eight to nine feet have either died young or been crippled early by failure of joints and bones. At half our size, we could not wield a club with sufficient force to hunt large animals (for kinetic energy would decrease 16 to 32-fold); we could not impart sufficient momentum to spears and arrows; we could not cut or split wood with primitive tools or mine minerals with picks and chisels. Since these all were essential activities in our historical development, we must conclude that the path of our evolution could only have been followed by a creature very close to our size. I do not argue that we inhabit the best of all possible worlds, only that our size has limited our activities and, to a great extent, shaped our evolution.

B | HUMAN BRAINS

An average human brain weighs about 1,300 grams (45.5 ounces); to accommodate such a large brain, we have bulbous, balloon-shaped heads unlike those of any other large mammal. Can we measure superiority by the size of our brains?

Elephants and whales have larger brains than ours. But this fact does not confer superior mental ability upon the largest mammals. Larger bodies need larger brains to coordinate their actions. We must find a way to remove the confusing influence of body size from our calculation. The computation of a simple ratio between brain weight and body weight will not work. Very small mammals generally have higher ratios than humans; that is, they have more brain per unit of body weight. Brain size does increase with body size, but it increases at a *much slower rate.*

If we plot brain weight against body weight for all species of adult mammals, we find that the brain increases at about two-thirds the rate of the body. Since surface areas also increase about two-thirds as fast as body weight, we conjecture that brain weight is not regulated by body weight, but primar-

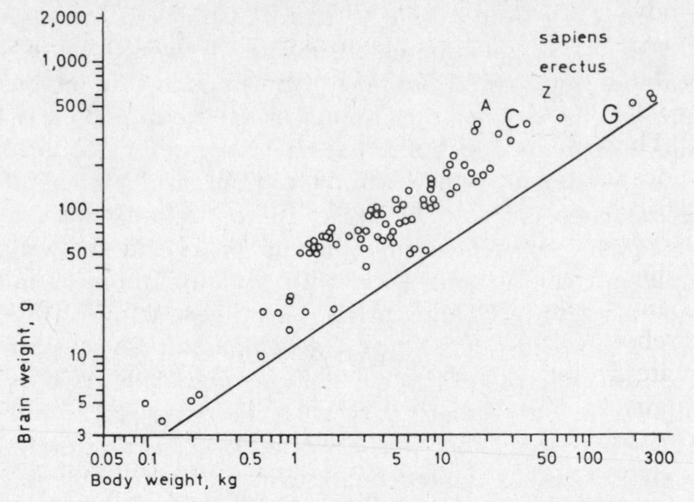

The correct criterion for assessing the superiority in size of our brains. The solid line represents the average relationship between brain weight and body weight for all body weights among mammals in general. Superiority in size is measured by upward deviation from this curve (i.e., "more" brain than an average mammal of the same body weight). Open circles represent primates (all have larger brains than average mammals). C is the chimpanzee, G the gorilla, and A the fossil hominid Australopithecus: *erectus covers the range of* Homo erectus *(Java and Peking Man); sapiens covers the field for modern humans. Our brains have the highest positive deviations of any mammal.* (F. S. Szalay, *Approaches to Primate Paleobiology,* Contrib. Primat. Vol. 5, 1975, p. 267. Reproduced with the permission of S. Karger AG, Basel)

ily by the body surfaces that serve as end points for so many innervations. This means that large animals may have absolutely larger brains than humans (because their bodies are bigger), and that small animals often have relatively larger brains than humans (because body size decreases more rapidly than brain size).

A plot of brain weight vs. body weight for adult mammals points the way out of our paradox. The correct criterion is neither absolute nor relative brain size—it is the difference between actual brain size and expected brain size at that body weight. To judge the size of our brain, we must compare it with the expected brain size for an average mammal of our body weight. On this criterion we are, as we had every right to expect, the brainiest mammal by far. No other species lies as far above the expected brain size for average mammals as we do.

This relationship between body weight and brain size provides important insights into the evolution of our brain. Our African ancestor (or at least close cousin), *Australopithecus africanus,* had an average adult cranial capacity of only 450 cubic centimeters. Gorillas often have larger brains, and many authorities have used this fact to infer a distinctly prehuman mentality for *Australopithecus.* A recent textbook states: "The original bipedal ape-man of South Africa had a brain scarcely larger than that of other apes and presumably possessed behavioral capacities to match." But *A. africanus* weighed only 50 to 90 pounds (female and male respectively—as estimated by Yale anthropologist David Pilbeam), while large male gorillas may weigh more than 600 pounds. We may safely state that *Australopithecus* had a much larger brain than other nonhuman primates, using the correct criterion of comparison with expected values for actual body weights.

The human brain is now about three times larger than that of *Australopithecus.* This increase has often been called the most rapid and most important event in the history of evolution. But our bodies have also increased greatly in size. Is this enlargement of the brain a simple consequence of bigger bodies or does it mark new levels of intelligence?

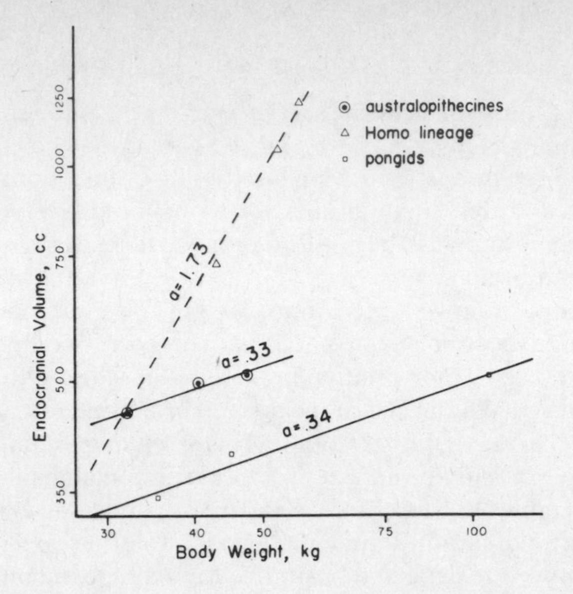

Evolutionary increase in human brain size (dotted line). The four triangles represent a rough evolutionary sequence: Australopithecus africanus, ER-1470 (Richard Leakey's new find with a cranial capacity just slightly less than 800 cc), Homo erectus *(Peking Man), and* Homo sapiens. *The slope is the highest ever calculated for an evolutionary sequence. The two solid lines represent more conventional scaling of brain size in australopithecines (above) and great apes (below).* ("Size and Scaling in Human Evolution," Pilbeam, David, and Gould, Stephen Jay, *Science* Vol. 186, pp. 892–901, Fig. 2, 6 December 1974. Copyright 1974 by the American Association for the Advancement of Science)

To answer this question, I have plotted cranial capacity against inferred body weight for the following fossil hominids (representing, perhaps, our lineage): *Australopithecus africanus;* Richard Leakey's remarkable find with a cranial capacity of nearly 800 cubic centimeters and an antiquity of more than two million years (weight estimated by David Pilbeam from dimensions of the femur); *Homo erectus* from Choukoutien (Peking Man); and modern *Homo sapiens.* The graph indicates that our brain has increased much more rapidly than any prediction based on compensations for body size would allow.

My conclusion is not unconventional, and it does reinforce

an ego that we would do well to deflate. Nonetheless, our brain has undergone a true increase in size not related to the demands of our larger body. We are, indeed, smarter than we were.

23 | History of the Vertebrate Brain

NATURE DISCLOSES the secrets of her past with the greatest reluctance. We paleontologists weave our tales from fossil fragments poorly preserved in incomplete sequences of sedimentary rocks. Most fossil mammals are known only from teeth—the hardiest substance in our bodies—and a few scattered bones. A famous paleontologist once remarked that mammalian history, as known from fossils, featured little more than the mating of teeth to produce slightly modified descendant teeth.

We rejoice at the rare preservation of soft parts—mammoths frozen in ice or insect wings preserved as carbonized films on beds of shale. Yet most of our information about the soft anatomy of fossils comes, not from these rare accidents, but from evidence commonly preserved in bone—the insertion scars of muscles or the holes through which nerves pass. Fortunately, the brain has also left its imprint upon the bones that enclose it. When a vertebrate dies, its brain quickly decays, but the resultant hole may be filled by sediment that hardens to produce a natural cast. This cast can preserve nothing of the brain's internal structure, but its size and external surface may faithfully copy the original.

Unfortunately, we cannot simply use the volume of a fossil cast as a reliable measure of an animal's intelligence; paleontology is never that easy. We must consider two problems.

First, what does brain size mean? Does it correlate at all with intelligence? There is no evidence for any relationship between intelligence and the normal range of variability for brain size *within* a species (fully functional human brains range from less than 1,000 to more than 2,000 cubic centimeters in volume). The variation among individuals within a species, however, is not the same phenomenon as variation in average values for different species. We must assume that, for example, average differences in brain size between humans and tuna fish bear some relationship to a meaningful concept of intelligence. Besides, what else can paleontologists do? We must work with what we have, and brain size is most of what we have.

Secondly, the primary determinant of brain size is not mental capacity, but body size. A large brain may reflect nothing more than the needs of the large body that housed it. Moreover, the relationship of brain size to body size is not a simple one (see previous essay). As animals get larger, brains increase in size at a slower rate. Small animals have relatively large brains; that is, the ratio of their brain weight to body weight is high. We must find some way to remove the influence of body size. This is done by plotting an equation for the "normal" relationship between brain weight and body weight.

Suppose we are studying mammals. We compile a list of average brain and body weights for adults of as many different species as we can. These species form the points of our graph; the equation fitting these points indicates that brain weight increases about two-thirds as fast as body weight. We can then compare the brain weight of any given species with the brain weight for an "average" mammal of that body weight. This comparison removes the influence of body size. A chimpanzee, for example, has an average brain weight of 395 grams. An average mammal of the same body weight should have a brain of 152 grams according to our equation. A chimp's brain is, therefore, 2.6 times as heavy as it "should" be (395/152). We may refer to this ratio of actual to expected brain size as an "encephalization quotient"; values greater than 1 signify larger than average brains; values less than 1 mark brains that are smaller than average.

But this method imposes another difficulty on paleontologists. We must now estimate body weight as well as brain weight. Complete skeletons are very rare and estimates are often made from a few major bones alone. To pile difficulty upon difficulty, only birds and mammals have brains that completely fill their cranial cavities. In these groups, a cranial cast faithfully reproduces the size and form of the brain. But in fishes, amphibians, and reptiles, the brain occupies only part of the cavity, and the fossilized cast is larger than the actual brain. We must estimate what part of the cast the brain would have occupied in life. And yet, despite this plethora of difficulties, assumptions, and estimates, we have been able to establish, and even to verify, a coherent and intriguing story about the evolution of brain size in vertebrates.

California psychologist Harry J. Jerison has recently marshaled all the evidence—much of it collected during his own labors of more than a decade—in a book entitled *The Evolution of the Brain and Intelligence* (New York, Academic Press, 1973).

Jerison's major theme is an attack upon the common idea that vertebrate classes can be arranged in a ladder of perfection leading from fish to mammal through the intermediary levels of amphibian, reptile, and bird. Jerison prefers a functional view that relates the amount of brain to specific requirements of modes of life, not to any preordained or intrinsic tendency for increase during the course of evolution. The potential "brain-body space" of modern vertebrates is filled in only two areas: one occupied by the warm-blooded vertebrates (birds and mammals), the other by their cold-blooded relatives (fish, amphibians, and modern reptiles). (Sharks provide the only exception to this general rule. Their brains are much too big—quite a surprise for these supposedly "primitive" fishes, but more on this later.) Warm-blooded vertebrates, to be sure, have larger brains than their cold-blooded relatives of the same body size, but there is no steady progress toward higher states, only a correlation between brain size and basic physiology. In fact, Jerison believes that mammals evolved their large brains to meet specific functional demands during their original existence as small creatures competing on the periphery of a world dominated by dinosaurs. He argues that the first mammals were nocturnal and that they needed larger brains to translate the perceptions of hearing and smell into spatial patterns that animals active in daylight could detect by vision alone.

Jerison presents a variety of intriguing tidbits within this framework. I hate to confute a comfortable item of received dogma, but I must report that dinosaurs did not have small brains—they had brains of just the right size for reptiles of their immense dimensions. We should never have expected more from *Brontosaurus* because large animals have relatively small brains, and reptiles, at any body weight, have smaller brains than mammals.

The gap between modern cold- and warm-blooded vertebrates is neatly filled by intermediate fossil forms. *Archaeopteryx,* the first bird, is known from fewer than half a dozen specimens, but one of them has a well-preserved brain cast. This intermediate form with feathers and reptilian teeth had a brain that plots right in the middle of the unfilled area between modern reptiles and birds. The primitive mammals that evolved so rapidly after dinosaurs became extinct had brains intermediate in size between reptiles and modern mammals of corresponding body weights.

We can even begin to understand the mechanism of this evolutionary increase in brain size by tracing one of the feedback loops that inspired it. Jerison computed the encephalization quotients for carnivores and their probable prey among ungulate herbivores for four separate groups: "archaic" mammals of the early Tertiary (the Tertiary is the conventional "age of mammals" and represents the last 70 million years of earth history); advanced mammals of the early Tertiary; middle to late Tertiary mammals; and modern mammals. Remember that an encephalization quotient of 1.0 denotes the expected brain size of an average modern mammal.

	Herbivores	Carnivores
Early Tertiary (archaic)	0.18	0.44
Early Tertiary (advanced)	0.38	0.61
Middle to late Tertiary	0.63	0.76
Modern	0.95	1.10

Both herbivores and carnivores displayed continual increase in brain size during their evolution, but at each stage, the carnivores were always ahead. Animals that make a living by catching rapidly moving prey seem to need bigger brains than plant eaters. And, as the brains of herbivores grew larger (presumably under intense selective pressure imposed by their carnivorous predators), the carnivores also evolved bigger brains to maintain the differential.

South America provides a natural experiment to test this claim. Until the Isthmus of Panama rose just a couple of million years ago, South America was an isolated island continent. Advanced carnivores never reached this island, and predatory roles were filled by marsupial carnivores with low encephalization quotients. Here, the herbivores display no increase in brain size through time. Their average encephalization quotient remained below 0.5 throughout the Tertiary; moreover, these native herbivores were quickly eliminated when advanced carnivores crossed the isthmus from North America. Again, brain size is a functional adaptation to modes of life, not a quantity with an inherent tendency to increase. When we document an increase, we can relate it to specific requirements of ecological roles. Thus, we should not be surprised that "primitive" sharks have such large brains; they are, after all, the top carnivores of the sea, and brain size reflects mode of life, not time of evolutionary origin. Likewise, carnivorous dinosaurs like *Allosaurus* and *Tyrannosaurus* had larger brains than herbivores like *Brontosaurus.*

But what about our preoccupation with ourselves: does anything about the history of vertebrates indicate why one peculiar species should be so brainy? Here's a closing item for thought. The most ancient brain cast of a primate belongs to a 55-million-year-old creature named *Tetonius homunculus.* Jerison has calculated its encephalization quotient as 0.68. This is, to be sure, only two-thirds the size of an average *living* mammal of the same body weight, but it is by far the largest brain of its time (making the usual correction for body weight); in fact, it is more than three times as large as an average mammal of its period. Primates have been ahead right from the start; our large brain is only an exaggeration of a pattern set at the beginning of the age of mammals. But why did such a large brain evolve in a group of small, primitive, tree-dwelling mammals, more similar to rats and shrews than to mammals conventionally judged as more advanced? And with this provocative query, I end, for we simply do not know the answer to one of the most important questions we can ask.

24 Planetary Sizes and Surfaces

CHARLES LYELL expressed in no uncertain terms the guiding concept of his geologic revolution. In 1829, he wrote to his colleague and scientific opponent Roderick Murchison:

> My work . . . will endeavor to establish the *principle of reasoning* in the science . . . that *no causes whatever* have from the earliest time to which we can look back, to the present, ever acted, but those now acting; and that they never acted with different degrees of energy from that which they now exert.

The doctrine of slow, stately, essentially uniform rates of change had a profound influence on nineteenth-century thought. Darwin adopted it thirty years later, and paleontologists ever since have been searching for cases of slow and steady evolution in the fossil record. But where did Lyell's preference for gradual change originate?

All cosmic generalizations have complex roots. In part, Lyell merely "discovered" his own political prejudices in nature—if the earth proclaims that change must proceed slowly and gradually, encumbered by the weight of events long past, then liberals might take comfort in a world increasingly threatened by social unrest. Nature, however, is not merely an empty stage upon which scientists display their prior preferences; nature also speaks back. Many of the forces that shape the surface of our planet do act slowly and continuously. Lyell could measure the accumulation of silt in river bottoms and the gradual erosion of hillslopes. Lyell's gradualism, while far too extreme in his formulation, does express a large part of the earth's history.

Our planet's gradual processes arise from the action of what my colleagues Frank Press and Raymond Siever call the external and internal heat engines of the earth. Our sun powers the external engine, but its influence depends upon the earth's atmosphere. Press and Siever write:

> Solar energy drives the atmosphere in a complex pattern of winds to give us our climates and weather, and it drives the ocean's circulation in a pattern that is coupled to the atmosphere. The water and gases of the oceans and atmosphere chemically react with the solid surface and physically transport material from one place to another.

Most of these processes work gradually, in a classic Lyellian manner; their large results are an accumulation of minute changes. Running water wears the land away; dunes march over deserts; waves destroy the coastline in some places, while currents transport sand to extend it elsewhere.

Heat derived from radioactive decay powers the internal engine. Some of its results—earthquakes and volcanic eruptions, for example—strike us as sudden and catastrophic, but the basic process, discovered only a decade ago, must be a source of joy for Lyell's shade. Internal heat puts the earth's surface in motion, driving the continents apart at minute rates of centimeters per year. This gradual motion, extended over 200 million years, has separated the single land of Pangaea into our present, widely dispersed continents.

Yet our earth is decidedly atypical among the other inner planets of our solar system: Mercury, Mars, and our own moon. (I exclude Venus because we know almost nothing about its surface; only one Russian probe has successfully penetrated its dense atmosphere to send back but two ambiguous photos. I also exclude Jupiter and the large planets beyond. They are so much larger and less dense than the inner planets that they belong to a very different class of cosmic bodies.) No geologist, no matter how strong his prior preferences, could have preached a doctrine of uniformity on the surface of any inner planet except the earth.

Craters made by meteoritic bombardment dominate the surfaces of Mars, Mercury, and our moon. Indeed, the surface of Mercury is little more than a field of tightly packed and superimposed craters. The moon's surface is divided into two major areas: densely cratered highlands and the more sparsely cratered maria ("seas" of basaltic lava). Lyellian gradualism, so applicable to our earth, cannot possibly describe the history of our planetary neighbors.

Consider, for example, our moon's history, as inferred from data collected during the Apollo missions and summarized by Columbia University geologist W. Ian Ridley: The moon's crust rigidified more than 4 billion years ago. By 3.9 billion years ago, the greatest period of meteoritic bombardment had ended, the mare basins had been excavated, and the major craters formed. Between 3.1 and 3.8 billion years ago, radioactively generated heat produced the basaltic lava that filled the mare basins. Then the generation of new heat failed to match its loss at the lunar surface and the crust rigidified again; by 3.1 billion years ago, the crust became too thick to permit the ascent of any more basalt, and activity at the lunar surface essentially ended. Since then, nothing much has happened beyond the very occasional impact of a large meteorite and the constant influx of very small ones.

We view the moon today much as it looked 3 billion years ago. It has no atmosphere to erode and recycle the material of its surface, and it cannot generate the internal heat to churn up and change its visage. The moon is not dead, but it is certainly quiescent. The concentration of moonquakes at 800–1,000 km below the surface suggests a rigid crust of this thickness, compared with 70 km or so for the earth's lithosphere. A partially molten zone may exist below the lunar crust, but it is too far down to influence the surface. The moon's surface is ancient, and its record tells the story of its catastrophes—massive meteorites and upwelling lava. Its early history was marked by violent change; its last 3 billion years by very little indeed.

Why is the earth so different from its neighbors in recording a history marked in large part by cumulative gradual processes, rather than ancient catastrophes? Readers might be temped to think that the answer lies in some complicated difference of composition. But all the inner planets are basically similar, so far as we can tell, in density and mineralogical content. I wish to argue that the difference arises from a disarmingly simple fact—*size itself, and nothing else:* the earth is a good deal larger than its neighborts.

Galileo first discussed the cardinal importance of size in determining the form and operation of all physical objects (see essays 21 and 22). As a basic fact of geometry, large bodies are not subject to the same balance of forces as small objects of the same shape (all planets are, necessarily, roughly spherical). Consider the ratio of surface to volume in two spheres of different radii. Surface is measured by a constant times the radius squared; volume by a different constant times the radius cubed. Hence, volumes increase faster than surfaces as objects of the same shape become larger.

I maintain that Lyell's insight is a contingent result of the earth's relatively low surface/volume ratio, not a general characteristic of all change, as he would have argued. We begin by assuming that the earth's early history did not differ much from that of its neighbors. At one time, our planet must have been scarred by abundant craters. But they were effaced billions of years ago, destroyed by the earth's two heat machines: churned up by the internal machine (uplifted in mountains, covered by lava, or buried in the depths of the earth by subduction at the descending borders of lithospheric plates) or quickly obliterated in atmospheric or fluvial erosion by the external machine.

These two heat machines operate only because the earth is large enough to possess a relatively small surface and large gravitational field. Mercury and the moon have neither atmosphere nor an active surface. The external machine requires an atmosphere for its work. Newton's equation relates the force of gravity directly to the mass of two bodies and inversely to the square of the distance separating them. To calculate the gravitational force holding a molecule of water vapor on the earth and moon, we need only consider the mass of the planet (since the mass of the molecule is constant) and the distance from the planet's surface to its center. As a planet gets larger, its mass increases as the cube of its radius, while the squared distance from surface to center is simply the radius squared. Hence, as a planet gets larger, its gravitational pull on an atmospheric particle increases as r^3/r^2 (where r is the planet's radius). On the moon and Mercury, this force is too small to hold an atmosphere; even the heaviest particles do not abide long. The earth's gravity is sufficiently strong to hold a large, permanent atmosphere, to act as a medium for its external heat machine.

Internal heat is generated radioactively over the volume of a planet. It is radiated out into space at a planet's surface. Small planets, with their high ratio of surface to volume, quickly lose their heat and solidify their outer layers to relatively great depths. Larger planets retain their heat and the mobility of their surfaces.

The ideal test for this hypothesis would be a planet of intermediate size, for we predict that such a body would display a mix of early catastrophes and gradual processes. Mars, obligingly, is just the right size, nicely intermediate between the earth and our moon or Mercury. About half the Martian surface is cratered; the rest reflects the activity of rather limited internal and external heat machines. Martian gravity is weak compared to that of the earth, but it is strong enough to hold a slight atmosphere (about 200 times thinner than ours). High winds course over the Martian surface and dune fields have been observed. The evidence for fluvial erosion is even more impressive, if somewhat mysterious, given the paucity of water vapor in the Martian atmosphere. (The mystery has been much alleviated by the discovery that Mars's polar caps are predominantly frozen water, not carbon dioxide, as previously conjectured. It also seems likely that a considerable amount of water lies frozen as permafrost in the Martian soil. Carl Sagan has shown me photos of relatively small craters with lobate extensions in all directions. It is hard to interpret these features as anything but liquefied mud, flowing away from the crater following localized melting of permafrost upon impact. They cannot be made of lava because the meteorites that formed the craters were too small to generate enough heat on impact to melt rock.)

Evidence for internal heat is also abundant (and rather spectacular), while some recent speculation plausibly links it with the processes that move the earth's plates. Mars has a volcanic province with giant mountains surpassing anything on earth. Olympus Mons has a base 500 km wide, a height of 8 km and a crater 70 km in diameter. The nearby Vallis Marineris dwarfs any canyon on earth: it is 120 km wide, 6 km deep and more than 5,000 km long.

Now, the speculation: Many geologists believe that the earth's plates are moved by plumes of heat and molten material rising from deep within the earth (perhaps even at the core-mantle boundary, 3,200 km below the surface). These plumes emerge at the surface at relatively fixed "hot spots," and the earth's plates ride over the plumes. The Hawaiian Islands, for example, are an essentially linear chain increasing in age toward the northwest. If the Pacific plate is slowly moving over a fixed plume, then the Hawaiian Islands might have formed one by one.

Mars, at its intermediate size, should be more dynamic than the moon, less so than the earth. The moon's crust is too thick to move at all; internal heat does not reach the surface. The earth's crust is thin enough to break into plates and move continuously. Suppose that the crust of Mars is thin enough to allow heat to rise, but too thick to break up and move extensively. Suppose also that plumes exist both on the earth and Mars. Giant Olympus Mons may represent the locus of a plume, rising under a crust that cannot move —Olympus Mons, if you will, may be like all the Hawaiis, piled one atop the other. The Vallis Marineris may represent an unsuccessful "try" at plate tectonics—the crust fractured, but could not move.

Science, at its best, is unifying. It strikes my intellectual fancy to learn that the principle regulating a fly on my ceiling also determines the uniqueness of our earth among the inner planets (flies, as small animals, have a high ratio of surface to volume; gravitational forces, acting upon volume, are not strong enough to overcome the strength of surface adhesion holding a fly's foot to the ceiling). Pascal once remarked, in planetary metaphor, that knowledge is like a sphere in space; the more we learn—that is, the larger the sphere—the greater our contact with the unknown (the planet's surface). True enough—but remember the principle of surfaces and volumes! The larger the sphere, the greater the ratio of known (volume) to unknown (surface). May absolutely increased ignorance continue to flourish with relatively increased knowledge.

7 Science in Society—A Historical View

25 On Heroes and Fools in Science

AS A ROMANTIC teen-ager, I believed that my future life as a scientist would be justified if I could discover a single new fact and add a brick to the bright temple of human knowledge. The conviction was noble enough; the metaphor was simply silly. Yet that metaphor still governs the attitude of many scientists toward their subject.

In the conventional model of scientific "progress," we begin in superstitious ignorance and move toward final truth by the successive accumulation of facts. In this smug perspective, the history of science contains little more than anecdotal interest—for it can only chronicle past errors and credit the bricklayers for discerning glimpses of final truth. It is as transparent as an old-fashioned melodrama: truth (as we perceive it today) is the only arbiter and the world of past scientists is divided into good guys who were right and bad guys who were wrong.

Historians of science have utterly discredited this model during the past decade. Science is not a heartless pursuit of objective information. It is a creative human activity, its geniuses acting more as artists than as information processors. Changes in theory are not simply the derivative results of new discoveries but the work of creative imagination influenced by contemporary social and political forces. We should not judge the past through anachronistic spectacles of our own convictions—designating as heroes the scientists whom we judge to be right by criteria that had nothing to do with their

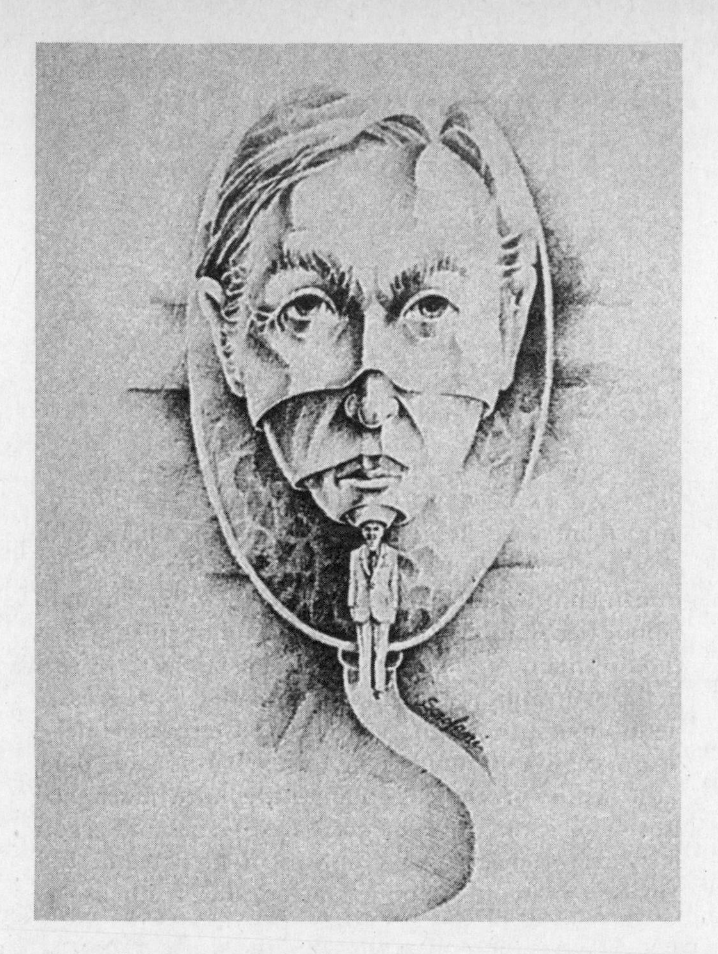

(Joseph Scrofani. Reproduced with permission, from *Natural History* Magazine, August-September 1974. © The American Museum of Natural History, 1974)

own concerns. We are simply foolish if we call Anaximander (sixth century B.C.) an evolutionist because, in advocating a primary role for water among the four elements, he held that life first inhabited the sea; yet most textbooks so credit him.

In this essay, I will take the most notorious of textbook baddies and try to display their theory as both reasonable in its time and enlightening in our own. Our villains are the eighteenth century "preformationists," adherents to an outmoded embryology. According to the textbooks, preformationists believed that a perfect miniature homunculus inhabited the human egg (or sperm), and that embryological development involved nothing more than its increase in size. The absurdity of this claim, the texts continue, is enhanced by its necessary corollary of *emboîtement* or encasement—for if Eve's ovum contained a homunculus, then the ovum of that homunculus contained a tinier homunculus, and so on into the inconceivable—a fully formed human smaller than an electron. The preformationists must have been blind, antiempirical dogmatists supporting an *a priori* doctrine of immutability against clear evidence of the senses—for one only has to open a chicken's egg in order to watch an embryo develop from simplicity to complexity. Indeed, their leading spokesman, Charles Bonnet, had proclaimed that "preformationism is the greatest triumph of reason over the senses." The heroes of our textbooks, on the other hand, were the "epigeneticists"; they spent their time looking at eggs rather than inventing fantasies. They proved by observation that the complexity of adult form developed gradually in the embryo. By the mid-nineteenth century, they had triumphed. One more victory for unsullied observation over prejudice and dogma.

In reality, the story is not so simple. The preformationists were as careful and accurate in their empirical observations as the epigeneticist. Moreover, if heroes we must have, that honor might as well fall to the preformationists who upheld, against the epigeneticists, a view of science quite congenial with our own.

The imagination of a few peripheral figures must not be taken as the belief of an entire school. The great preformationists—Malpighi, Bonnet, and von Haller,—all knew perfectly well that the chick embryo seemed to begin as a simple tube and become more and more complex as organs differentiated within the egg. They had studied and drawn the embryology of the chick in a series of astute observations that matched anything achieved by contemporary epigeneticists.

Preformationists and epigeneticists did not disagree about their observations; but, whereas epigeneticists were prepared to take those observations literally, the preformationists insisted on probing "behind appearance." They claimed that the visual manifestations of development were deceptive. The early embryo is so tiny, so gelatinous, and so transparent that the preformed structures could not be discerned by the crude microscopes then available. Bonnet wrote in 1762: "Do not mark the time when organized beings begin to exist by the time when they begin to become visible; and do not constrain nature by the strict limits of our senses and instruments." Moreover, the preformationists never believed that preformed structures were organized into a perfect miniature homunculus in the egg itself. The rudiments existed in the egg to be sure, but in relative positions and proportions bearing little relationship to adult morphology. Again, Bonnet in 1762: "While the chick is still a germ, all its parts have forms, proportions and positions which differ greatly from those that they will attain during development. If we were able to see the germ enlarged, as it is when small, it would be impossible for us to recognize it as a chick. All the parts of the germ do not develop at the same time and uniformly."

But how did the preformationists explain the *reductio ad absurdum* of encasement—the encapsulation of our entire history in the ovaries of Eve? Very simply—this concept was not absurd in an eighteenth-century context.

First of all, scientists believed that the world had existed—and would endure—for only a few thousand years. One had, therefore, to encapsulate only a limited number of generations, not the potential products of several million years on a twentieth-century geological time chart.

Secondly, the eighteenth century had no cell theory to set a lower boundary to organic size. It now seems absurd to postulate a fully formed homunculus smaller than the minimum size of a single cell. But an eighteenth-century scientist had no reason to postulate a lower limit to size. In fact, it was widely believed that Leeuwenhoek's animalcules, the single-celled microscopic creatures that had so aroused the imagination of Europe, had complete sets of miniature organs. Thus Bonnet, supporting the corpuscular theory (that light is made of discrete particles), rhapsodized about the inconceivable tininess of the several million globules of light that penetrate all at once into the supposed eyes of animalcules. "Nature works as small as it wishes. We know not at all the lower boundary of the division of matter, but we see that it has been prodigiously divided. From the elephant to the mite, from the whale to the animalcule 27 million times smaller than the mite, from the globe of the sun to the globule of light, what an inconceivable multitude of intermediate degrees!"

Why did the preformationists feel such a need to penetrate behind appearances? Why would they not accept the direct evidence of their senses? Consider the alternatives. Either the parts are present from the first or the fertilized egg is utterly formless. If the egg is formless, then some external force must unerringly impose a design upon matter only potentially capable of producing it. But what kind of a force could this be? And must there be a different force for each species of animal? How can we learn about it, test it, perceive it, touch it, or understand it? How could it represent any more than an insubstantial appeal to a mysterious and mystical vitalism?

Preformationism represented the best of Newtonian science. It was designed to save a general attitude, which we would recognize today as "scientific," from a vitalism that the evidence of raw sensation implied. If the egg were truly unorganized, homogeneous material without preformed parts, then how could it yield such wondrous complexity without a mysterious directing force? It does so, and can do so, only because the structure (not merely the raw material) needed to build this complexity already resides in the egg. In this light, Bonnet's statement about the triumph of reason over the senses seems itself more reasonable.

Finally, who can say that our current understanding of embryology marks the triumph of epigenesis? Most great debates are resolved at Aristotle's golden mean, and this is no exception. From our perspective today, the epigeneticists were right; organs differentiate sequentially from simpler rudiments during embryological development; there are no preformed parts. But the preformationists were also right in insisting that complexity cannot arise from formless raw material—that there must be something within the egg to regulate its development. All we can say (as if it mattered) is that they incorrectly identified this "something" as preformed parts, where we now understand it as encoded instructions built of DNA. But what else could we expect from eighteenth-century scientists, who knew nothing of the player piano, not to mention the computer program? The idea of a coded program was not part of their intellectual equipment.

And, come to think of it, what could be more fantastic than the claim that an egg contains thousands of instructions, written on molecules that tell the cell to turn on and off the production of certain substances that regulate the speed of chemical processes? The notion of preformed parts sounds far less contrived to me. The only thing going for coded instructions is that they seem to be there.

26 Posture Maketh the Man

NO EVENT DID MORE to establish the fame and prestige of The American Museum of Natural History than the Gobi Desert expeditions of the 1920s. The discoveries, including the first dinosaur eggs, were exciting and abundant, and the sheer romance fit Hollywood's most heroic mold. It is still hard to find a better adventure story than Roy Chapman Andrews's book (with its chauvinistic title): *The New Conquest of Central Asia.* Nonetheless, the expeditions utterly failed to achieve their stated purpose: to find in Central Asia the ancestors of man. And they failed for the most elementary of reasons—we evolved in Africa, as Darwin had surmised fifty years earlier.

Our African ancestors (or at least our nearest cousins) were discovered in cave deposits during the 1920s. But these australopithecines failed to fit preconceived notions of what a "missing link" should look like, and many scientists refused to accept them as bona fide members of our lineage. Most anthropologists had imagined a fairly harmonious transformation from ape to human, propelled by increasing intelligence. A missing link should be intermediate in both body and brain—Alley Oop or the old (and false) representations of stoopshouldered Neanderthals. But the australopithecines refused to conform. To be sure, their brains were bigger than those of any ape with comparable body size, but not much bigger (see essays 22 and 23). Most of our evolutionary increase in brain size occurred after we reached the australopithecine level. Yet these smallbrained australopithecines walked as erect as you or I. How could this be? If our evolution was propelled by an enlarging brain, how could upright posture—another "hallmark of hominization," not just an incidental feature—originate first? In a 1963 essay, George Gaylord Simpson used this dilemma to illustrate

> the sometimes spectacular failure to predict discoveries even when there is a sound basis for such prediction. An evolutionary example is the failure to predict discovery of a "missing link," now known [*Australopithecus*], that was upright and tool-making but had the physiognomy and cranial capacity of an ape.

We must ascribe this "spectacular failure" primarily to a subtle prejudice that led to the following, invalid extrapolation: We dominate other animals by brain power (and little else); therefore, an increasing brain must have propelled our own evolution at all stages. The tradition for subordinating upright posture to an enlarging brain can be traced throughout the history of anthropology. Karl Ernst von Baer, the greatest embryologist of the nineteenth century (and second only to Darwin in my personal pantheon of scientific heroes) wrote in 1828: "Upright posture is only the consequence of the higher development of the brain . . . all differences between men and other animals depend upon construction of the brain." One hundred years later, the English anthropologist G. E. Smith wrote: "It was not the adoption of the erect attitude or the invention of articulate language that made man from an ape, but the gradual perfecting of a brain and the slow building of the mental structure, of which erectness of carriage and speech are some of the incidental manifestations."

Against this chorus of emphasis upon the brain, a very few scientists upheld the primacy of upright posture. Sigmund Freud based much of his highly idiosyncratic theory for the origin of civilization upon it. Beginning in his letters to Wilhelm Fliess in the 1890s and culminating in his 1930 essay on *Civilization and Its Discontents,* Freud argued that our assumption of upright posture had reoriented our primary sensation from smell to vision. This devaluation of olfaction shifted the object of sexual stimulation in males from cyclic odors of estrus to the continual visibility of female genitalia. Continual desire of males led to the evolution of continual receptivity in females. Most mammals copulate only around periods of ovulation; humans are sexually active at all times (a favorite theme of writers on sexuality). Continual sexuality has cemented the human family and made civilization possible; animals with strongly cyclic copulation have no strong impetus for stable family structure. "The fateful process of civilization," Freud concludes, "would thus have set in with man's adoption of an erect posture."

Although Freud's ideas gained no following among anthropologists, another minor tradition did arise to stress the primacy of upright posture. (It is, by the way, the argument we tend to accept today in explaining the morphology of australopithecines and the path of human evolution.) The brain cannot begin to increase in a vacuum. A primary impetus must be provided by an altered mode of life that would place a strong, selective premium upon intelligence. Upright posture frees the hands from locomotion and for manipulation (literally, from *manus* = "hand"). For the first time, tools and weapons can be fashioned and used with ease. Increased intelligence is largely a response to the enormous potential inherent in free hands for manufacture—again, literally. (Needless to say, no anthropologist has ever been so naïve as to argue that brain and posture are completely independent in evolution, that one reached its fully human status before the other began to change at all. We are dealing with interaction and mutual reinforcement. Nevertheless, our early evolution did involve a more rapid change in posture than in brain size; complete freeing of our hands for using tools preceded most of the evolutionary enlargement of our brain).

In another proof that sobriety does not make right, von Baer's mystical and oracular colleague Lorenz Oken hit upon the "correct" argument in 1809, while von Baer was led astray a few years later. "Man by the upright walk obtains his character," writes Oken, "the hands become free and can achieve all other offices. . . . With the freedom of the body has been granted also the freedom of the mind." But the champion of upright posture during the nineteenth century was Darwin's German bulldog Ernst Haeckel. Without a scrap of direct evidence, Haeckel reconstructed our ancestor and even gave it a scientific name, *Pithecanthropus alalus,* the upright, speechless, small-brained ape-man. (*Pithecanthropus,* by the way, is probably the only scientific name ever given to an animal before it was discovered. When Du Bois discovered Java Man in the 1890s, he adopted Haeckel's generic name but he gave it the new specific designation *Pithecanthropus erectus.* We now usually include this creature in our own genus as *Homo erectus.*)

But why, despite Oken and Haeckel's demurral, did the idea of cerebral primary become so strongly entrenched? One thing is sure; it had nothing to do with direct evidence —for there was none for any position. With the exception of Neanderthal (a geographic variant of our own species according to most anthropologists), no fossil humans were discovered until the closing years of the nineteenth century, long after the dogma of cerebral primary was established. But debates based on no evidence are among the most revealing in the history of science, for in the absence of factual constraints, the cultural biases that affect all thought (and which scientists try so assiduously to deny) lie nakedly exposed.

Indeed, the nineteenth century produced a brilliant exposé from a source that will no doubt surprise most readers— Friedrich Engels. (A bit of reflection should diminish surprise. Engels had a keen interest in the natural sciences and sought to base his general philosophy of dialectical materialism upon a "positive" foundation. He did not live to complete his "dialectics of nature," but he included long commentaries on science in such treatises as the *Anti-Dühring.*) In 1876, Engels wrote an essay entitled, *The Part Played by Labor in the Transition from Ape to Man.* It was published posthumously in 1896 and, unfortunately, had no visible impact upon Western science.

Engels considers three essential features of human evolution: speech, a large brain, and upright posture. He argues that the first step must have been a descent from the trees with subsequent evolution to upright posture by our ground-dwelling ancestors. "These apes when moving on level ground began to drop the habit of using their hands and to adopt a more and more erect gait. This was the decisive step in the transition from ape to man." Upright posture freed the hand for using tools (labor, in Engels's terminology); increased intelligence and speech came later.

> Thus the hand is not only the organ of labor, it is also the product of labor. Only by labor, by adaptation to ever new operations . . . by the ever-renewed employment of these inherited improvements in new, more and more complicated operations, has the human hand attained the high degree of perfection that has enabled it to conjure into being the pictures of Raphael, the statues of Thorwaldsen, the music of Paganini.

Engels presents his conclusions as though they followed deductively from the premises of his materialist philosophy, but I am confident that he cribbed them from Haeckel. The two formulations are almost identical, and Engels cites the relevant pages of Haeckel's work for other purposes in an earlier essay written in 1874. But no matter. The importance of Engels's essay lies, not in its substantive conclusions, but in its trenchant political analysis of why Western science was so hung up on the a priori assertion of cerebral primacy.

As humans learned to master their material surroundings, Engels argues, other skills were added to primitive hunting—agriculture, spinning, pottery, navigation, arts and sciences, law and politics, and finally, "the fantastic reflection of human things in the human mind: religion." As wealth accumulated, small groups of men seized power and forced others to work for them. Labor, the source of all wealth and the primary impetus for human evolution, assumed the same low status of those who labored for the rulers. Since rulers governed by their will (that is, by feats of mind), actions of the brain appeared to have a motive power of their own. The profession of philosophy followed no unsullied ideal of truth.

Philosophers relied on state or religious patronage. Even if Plato did not consciously conspire to bolster the privileges of rulers with a supposedly abstract philosophy, his own class position encouraged an emphasis on thought as primary, dominating, and altogether more noble and important than the labor it supervised. This idealistic tradition dominated philosophy right through to Darwin's day. Its influence was so subtle and pervasive that even scientific, but apolitical, materialists like Darwin fell under its sway. A bias must be recognized before it can be challenged. Cerebral primacy seemed so obvious and natural that it was accepted as given, rather than recognized as a deep-seated social prejudice related to the class position of professional thinkers and their patrons. Engels writes:

> All merit for the swift advance of civilization was ascribed to the mind, to the development and activity of the brain. Men became accustomed to explain their actions from their thoughts, instead of from their needs. . . . And so there arose in the course of time that idealistic outlook on the world which, especially since the downfall of the ancient world, has dominated men's minds. It still rules them to such a degree that even the most materialistic natural scientists of the Darwinian school are still unable to form any clear idea of the origin of man, because under that ideological influence they do not recognize the part that has been played therein by labor.

The importance of Engels's essay does not lie in the happy result that *Australopithecus* confirmed a specific theory proposed by him—via Haeckel—but rather in his perceptive analysis of the political role of science and of the social biases that must affect all thought.

Indeed, Engels's theme of the separation of head and hand has done much to set and limit the course of science throughout history. Academic science, in particular, has been constrained by an ideal of "pure" research, which in former days barred a scientist from extensive experimentation and empirical testing. Ancient Greek science labored under the restriction that patrician thinkers could not perform the manual work of plebeian artisans. Medieval barber-surgeons who had to deal with battlefield casualties did more to advance the practice of medicine than academic physicians who rarely examined patients and who based their treatment on a knowledge of Galen and other learned texts. Even today, "pure" researchers tend to disparage the practical, and terms such as "aggie school" and "cow college" are heard with distressing frequency in academic circles. If we took Engels's message to heart and recognized our belief in the inherent superiority of pure research for what it is—namely social prejudice—then we might forge among scientists the union between theory and practice that a world teetering dangerously near the brink so desperately needs.

27 Racism and Recapitulation

The adult who retains the more numerous fetal, [or] infantile . . . traits is unquestionably inferior to him whose development has progressed beyond them. Measured by these criteria, the European or white race stands at the head of the list, the African or negro at its foot.

D.G. BRINTON, *1890*

On the basis of my theory, I am obviously a believer in the inequality of races. . . . In his fetal development the negro passes through a stage that has already become the final stage for the white man. If retardation continues in the negro, what is still a transitional stage may for this race also become a final one. It is possible for all other races to reach the zenith of development now occupied by the white race.

L. BOLK, *1926*

BLACKS ARE inferior, Brinton tells us, because they retain juvenile traits. Blacks are inferior, claims Bolk, because they develop beyond the juvenile traits that whites retain. I doubt that anyone could construct two more contradictory arguments to support the same opinion.

The arguments arise from different readings of a fairly technical subject in evolutionary theory: the relationship between ontogeny (the growth of individuals) and phylogeny (the evolutionary history of lineages). My aim here is not to explicate this subject but rather to make a point about pseu-

The 1874 edition of Ernst Haeckel's Anthropogenie *contains this racist illustration of evolution. (Courtesy of the American Museum of Natural History)*

doscientific racism. We like to think that scientific progress drives out superstition and prejudice. Brinton linked his racism to the theory of recapitulation, the belief that individuals, in their own embryonic and juvenile growth, repeat the adult stages of their ancestors—that each individual, in its own development, climbs up its family tree. (To supporters of recapitulation, the embryonic gill slits of human fetuses represent the adult fish from which we descended. And, in the racist reading, white children will pass through and beyond the intellectual stages that characterize adults of "lower" races.) During the late nineteenth century, recapitulation provided one of the two or three leading "scientific" arguments in the racist arsenal.

By the end of the 1920s, however, the theory of recapitulation had utterly collapsed. In fact, as I argue in essay 7, anthropologists began to interpret human evolution in precisely the opposite manner. Bolk led the movement, arguing that humans evolved by retaining the juvenile stages of our ancestors and losing previously adult structures—a process called neoteny. With this reversal, we might have expected a rout of white racism: at least, a quiet putting aside of previous claims; at best, an honest admission that the old evidence, interpreted under the new theory of neoteny, affirmed the superiority of blacks (since the retention of juvenile features now becomes a progressive trait). No such thing happened. The old evidence was quietly forgotten, and Bolk sought new data to contradict the old information and support once again the inferiority of blacks. With neoteny, "higher" races must retain more juvenile traits as adults; so Bolk discarded all the embarrassing "facts" once used by recapitulationists and enlisted the few juvenile features of adult whites in his support.

Clearly, science did not influence racial attitudes in this case. Quite the reverse: an a priori belief in black inferiority determined the biased selection of "evidence." From a rich body of data that could support almost any racial assertion, scientists selected facts that would yield their favored conclusion according to theories currently in vogue. There is, I believe, a general message in this sad tale. There is not now and there never has been any unambiguous evidence for genetic determination of traits that tempt us to make racist distinctions (differences between races in average values for brain size, intelligence, moral discernment, and so on). Yet this lack of evidence has not forestalled the expression of scientific opinion. We must therefore conclude that this expression is a political rather than a scientific act—and that scientists tend to behave in a conservative way by providing "objectivity" for what society at large wants to hear.

To return to my story: Ernst Haeckel, Darwin's greatest popularizer, saw great promise for evolutionary theory as a social weapon. He wrote:

> Evolution and progress stand on the one side, marshaled under the bright banner of science; on the other side, marshaled under the black flag of hierarchy, stand spiritual servitude and falsehood, want of reason and barbarism, superstition and retrogression. . . . Evolution is the heavy artillery in the struggle for truth; whole ranks of dualistic sophisms fall before [it] . . . as before the chain shot of artillery.

Recapitulation was Haeckel's favorite argument (he named it the "biogenetic law" and coined the phrase "ontogeny recapitulates phylogeny"). He used it to attack nobility's claim to special status—are we not all fish as embryos?—and to ridicule the soul's immortality—for where could the soul be in our embryonic, wormlike condition?

Haeckel and his colleagues also invoked recapitulation to affirm the racial superiority of northern European whites. They scoured the evidence of human anatomy and behavior, using everything they could find from brains to belly buttons. Herbert Spencer wrote that "the intellectual traits of the uncivilized . . . are traits recurring in the children of the civilized." Carl Vogt said it more strongly in 1864: "The grown up Negro partakes, as regards his intellectual faculties, of the nature of the child. . . . Some tribes have founded states, possessing a peculiar organization, but, as to the rest, we may boldly assert that the whole race has, neither in the past nor in the present, performed anything tending to the progress of humanity or worthy of preservation." And the French medical anatomist Etienne Serres really did argue that black males are primitive because the distance between their navel and penis remains small (relative to body height) throughout life, while white children begin with a small separation but increase it during growth—the rising belly button as a mark of progress.

The general argument found many social uses. Edward Drinker Cope, best known for his "fossil feud" with Othniel Charles Marsh, compared the cave art of Stone Age man with that of white children and "primitive" adults living today: "We find that the efforts of the earliest races of which we have any knowledge were similar to those which the untaught hand of infancy traces on its slate or the savage depicts on the rocky faces of hills." A whole school of "criminal anthropology" (see next essay) branded white wrongdoers as genetically retarded and compared them again with children and adult Africans or Indians: "Some of them [white criminals]," wrote one zealous supporter, "would have been the ornament and moral aristocracy of a tribe of Red Indians." Havelock Ellis noted that white criminals, white children, and South American Indians generally do not blush.

Recapitulation had its greatest political impact as an argument to justify imperialism. Kipling, in his poem on the "white man's burden," referred to vanquished natives as "half devil and half child." If the conquest of distant lands upset some Christian beliefs, science could always relieve a bothered conscience by pointing out that primitive people, like white children, were incapable of self-government in a modern world. During the Spanish-American War, a major debate arose in the United States over whether we had a right to annex the Philippines. When antiimperialists cited Henry Clay's contention that the Lord would not have created a race incapable of self-government, Rev. Josiah Strong replied: "Clay's conception was formed before modern science had shown that races develop in the course of centuries as individuals do in years, and that an underdeveloped race, which is incapable of self-government, is no more of a reflection on the Almighty than is an undeveloped child who is incapable of self-government." Others took the "liberal" viewpoint and cast their racism in the paternalist mode: "Without primitive peoples, the world at large would be much what in small it is without the blessing of children. . . . We ought to be as fair to the 'naughty race' abroad as we are to the 'naughty boy' at home."

But the theory of recapitulation contained a fatal flaw. If the adult traits of ancestors become juvenile features of descendants, then their development must be speeded up to make room for the addition of new adult characters onto the end of a descendant's ontogeny. With the rediscovery of Mendelian genetics in 1900, this "law of acceleration" collapsed, carrying with it the whole theory of recapitulation—for if genes make enzymes and enzymes control the rates of processes, then evolution may act either by speeding up or slowing down the rate of development. Recapitulation requires a universal speeding up, but genetics proclaims that slowing down is just as likely. When scientists began to look for evidences of slowing down, our own species took the limelight. As I argue in essay 7, humans have, in many respects, evolved by retaining juvenile features common to primates and even to mammals in general—for example, our bulbous cranium and relatively large brain, the ventral position of our foramen magnum (permitting upright posture), small jaws, and relative hairlessness.

For a half century the proponents of recapitulation had collected racial "evidence"; all of it argued that adults of "lower" races were like white children. When the theory of recapitulation collapsed, supporters of human neoteny still had these data. An objective reinterpretation should have led to an admission that "lower" races are superior; for as Havelock Ellis (an early supporter of neoteny) wrote: "The progress of our race has been a progress in youthfulness." Indeed, the new criterion was accepted—the more childlike race would henceforward wear the mantle of superiority. But the old evidence was simply discarded, and Bolk scurried about for some opposing information to prove that adult whites are like black children. He found it, of course (you always can if you want to badly enough): adult blacks have long skulls, dark skins, strongly prognathous jaws, and an "ancestral dentition"; while adult whites and black babies have short skulls, light (or at least lighter) skins, and small, nonjutting jaws (we'll pass on the teeth). "The white race appears to be the most progressive, as being the most retarded," said Bolk. Havelock Ellis had said much the same in 1894: "The child of many African races is scarcely if at all less intelligent than the European child, but while the African as he grows up becomes stupid and obtuse, and his whole social life falls into a state of hidebound routine, the European retains much of his childlike vivacity."

Lest we dismiss these statements as lapses of a bygone age, I note that the neotenic argument was invoked in 1971 by a leading genetic determinist in the IQ debate. H. Eysenck claims that African and black American babies display faster sensorimotor development than whites. He also argues that rapid sensorimotor development in the first year of life correlates with lower IQ later. This is a classic example of a potentially meaningless, noncausal correlation: suppose that differences in IQ are completely determined by environment; then, rapid motor development does not cause low IQ—it is merely another measure of racial identification (and a poorer one than skin color). Nonetheless, Eysenck invokes neoteny to support his genetic interpretation: "These findings are important because of a very general view in biology according to which the more prolonged the infancy the greater in general are the cognitive or intellectual abilities of the species."

But there is a hooker in the neotenic argument, one that white racists have generally chosen to ignore. It can scarcely be denied that the most juvenilized of human races are not white, but mongoloid (something the American military never understood when it claimed that the Vietcong were manning their armies with "teen-agers"—many of whom turned out to be in their thirties or forties). Bolk darted around it; Havelock Ellis met it squarely and admitted defeat (if not inferiority).

If the racist recapitulationists lost their theory, perhaps the racist neotenists will lose on facts (even though history suggests that facts are simply selected to fit prior theories). For there is another embarrassing point in the data of neoteny—namely, the status of women. All was well under recapitulation. Women are more childlike in their anatomy than men—a sure sign of inferiority, as Cope argued so vociferously in the 1880s. Yet, in the neotenic hypothesis, women should be superior by the same evidence. Again, Bolk chose to ignore the issue. And again, Havelock Ellis met it honestly to admit the position that Ashley Montagu later championed in his treatise on "the natural superiority of women." Ellis wrote in 1894: "She bears the special characteristics of humanity in a higher degree than man. . . . This is true of physical characters: the largeheaded, delicate-faced, small-boned man of urban civilization is much nearer to the typical woman than is the savage. Not only by his large brain, but by his large pelvis, the modern man is following a path first marked out by woman." Ellis even suggested that we might seek our salvation in the closing lines of Faust:

> Eternal womanhood
> Lead us on high.

28 The Criminal as Nature's Mistake, or the Ape in Some of Us

W.S. GILBERT DIRECTED his potent satire at all forms of pretension as he saw them. For the most part we continue to applaud him: pompous peers and affected poets are still legitimate targets. But Gilbert was a comfortable Victorian at heart, and much that he labeled as pretentious now strikes us as enlightened—higher education for women, in particular.

A women's college! maddest folly going!
What can girls learn within its walls worth knowing?

In *Princess Ida,* the Professor of Humanities at Castle Adamant provides a biological justification for her proposition that "man is nature's sole mistake." She tells the tale of an ape who loved a beautiful woman. To win her affection, he tried to dress and act like a gentleman, but all necessarily in vain, for

Darwinian Man, though well-behaved,
At best is only a monkey shaved!

Gilbert produced *Princess Ida* in 1884, eight years after an Italian physician, Cesare Lombroso, had initiated one of the most powerful social movements of his time with a similar claim made in all seriousness about a group of men—born criminals are essentially apes living in our midst. Later in life, Lombroso recalled his moment of revelation:

> In 1870 I was carrying on for several months researches in the prisons and asylums of Pavia upon cadavers and living persons, in order to determine upon substantial differences between the insane and criminals, without succeeding very well. Suddenly, the morning of a gloomy day in December, I found in the skull of a brigand a very long series of atavistic anomalies. . . . The problem of the nature and of the origin of the criminal seemed to me resolved; the characters of primitive men and of inferior animals must be reproduced in our times.

Biological theories of criminality were scarcely new, but Lombroso gave the argument a novel, evolutionary twist. Born criminals are not simply deranged or diseased; they are, literally, throwbacks to a previous evolutionary stage. The hereditary characters of our primitive and apish ancestors remain in our genetic repertoire. Some unfortunate men are born with an unusually large number of these ancestral characters. Their behavior may have been appropriate in savage societies of the past; today, we brand it as criminal. We may pity the born criminal, for he cannot help himself; but we cannot tolerate his actions. (Lombroso believed that about 40 percent of criminals fell into this category of innate biology—born criminals. Others committed misdeeds from greed, jealousy, extreme anger, and so on—criminals of occasion.)

I tell this tale for three reasons that combine to make it far more than an antiquarian exercise in a small corner of forgotten, late-nineteenth-century history.

1. A generalization about social history: It illustrates the enormous influence of evolutionary theory in fields far removed from its biological core. Even the most abstract scientists are not free agents. Major ideas have remarkably subtle and far-ranging extensions. The inhabitants of a nuclear world should know this perfectly well, but many scientists have yet to get the message.

2. A political point: Appeals to innate biology for the explanation of human behavior have often been advanced in the name of enlightenment. The proponents of biological determinism argue that science can cut through a web of supersti-

tion and sentimentalism to instruct us about our true nature. But their claims have usually had a different primary effect: they are used by the leaders of class-stratified societies to assert that a current social order must prevail because it is the law of nature. Of course, no view should be rejected because we dislike its implications. Truth, as we understand it, must be the primary criterion. But the claims of determinists have always turned out to be prejudiced speculation, not ascertained fact—and Lombroso's criminal anthropology is the finest example I know.

3 . A contemporary note: Lombroso's brand of criminal anthropology is dead, but its basic postulate lives on in popular notions of criminal genes or chromosomes. These modern incarnations are worth about as much as Lombroso's original version. Their hold on our attention only illustrates the unfortunate appeal of biological determinism in our continuing attempt to exonerate a society in which so many of us flourish by blaming the victim.

The year 1976 marked the centenary of Lombroso's founding document—later enlarged into the famous *L'uomo delinquente (Criminal Man).* Lombroso begins with a series of anecdotes to assert that the usual behavior of lower animals is criminal by our standards. Animals murder to suppress revolts; they eliminate sexual rivals; they kill from rage (an ant, made impatient by a recalcitrant aphid, killed and devoured it); they form criminal associations (three communal beavers shared a territory with a solitary individual; the trio visited their neighbor and were well treated; when the loner returned the visit, he was killed for his solicitude). Lombroso even brands the fly catching of insectivorous plants as an "equivalent of crime" (although I fail to see how it differs from any other form of eating).

In the next section, Lombroso examines the anatomy of criminals and finds the physical signs (stigmata) of their primitive status as throwbacks to our evolutionary past. Since he has already defined the normal behavior of animals as criminal, the actions of these living primitives must arise from their nature. The apish features of born criminals include relatively long arms, prehensile feet with mobile big toes, low and narrow forehead, large ears, thick skull, large and prognathous jaw, copious hair on the male chest, and diminished sensitivity to pain. But the throwbacks do not stop at the primate level. Large canine teeth and a flat palate recall a more distant mammalian past. Lombroso even compares the heightened facial asymmetry of born criminals with the normal condition of flatfishes (both eyes on one side of the head)!

But the stigmata are not only physical. The social behavior of the born criminal also allies him with apes and living human savages. Lombroso placed special emphasis on tattooing, a common practice among primitive tribes and European criminals. He produced voluminous statistics on the content of criminal tattoos and found them lewd, lawless, or exculpating (although one read, he had to admit, *Vive la France et les pommes de terres frites*—"long live France and french fried potatoes"). In criminal slang, he found a language of its own, markedly similar to the speech of savage tribes in such features as onomatopoeia and personification of inanimate objects: "They speak differently because they feel differently; they speak like savages, because they are true savages in the midst of our brilliant European civilization."

Lombroso's theory was no work of abstract science. He founded and actively led an international school of "criminal anthropology" that spearheaded one of the most influential of late-nineteenth-century social movements. Lombroso's "positive," or "new," school campaigned vigorously for changes in law enforcement and penal practices. They regarded their improved criteria for the recognition of born criminals as a primary contribution to law enforcement. Lombroso even suggested a preventive criminology—society need not wait (and suffer) for the act itself, for physical and social stigmata define the potential criminal. He can be identified (in early childhood), watched, and whisked away at the first manifestation of his irrevocable nature (Lombroso, a liberal, favored exile rather than death). Enrico Ferri, Lombroso's closest colleague, recommended that "tattooing, anthropometry, physiognomy . . . reflex activity, vasomotor reactions [criminals, he argued, do not blush], and the range of sight" be used as criteria of judgment by magistrates.

Criminal anthropologists also campaigned for a basic reform in penal practice. An antiquated Christian ethic held that criminals should be sentenced for their deeds, but biology declares that they should be judged by their nature. Fit the punishment to the criminal, not to the crime. Criminals of occasion, lacking the stigmata and capable of reform, should be jailed for the term necessary to secure their amendment. But born criminals are condemned by their nature: "Theoretical ethics passes over the diseased brain, as oil does over marble, without penetrating it." Lombroso recommended irrevocable detention for life (in pleasant, but isolated surroundings) for any recidivist with the telltale stigmata. Some of his colleagues were less generous. An influential jurist wrote to Lombroso:

> You have shown us fierce and lubricious orang-utans with human faces. It is evident that as such they cannot act otherwise. If they ravish, steal, and kill, it is by virtue of their own nature and their past, but there is all the more reason for destroying them when it has been proved that they will always remain orang-utans.

And Lombroso himself did not rule out the "final solution":

> The fact that there exist such beings as born criminals, organically fitted for evil, atavistic reproductions, not simply of savage men but even of the fiercest animals, far from making us more compassionate towards them, as has been maintained, steels us against all pity.

One other social impact of Lombroso's school should be mentioned. If human savages, like born criminals, retained apish traits, then primitive tribes—"lesser breeds without the law"—could be regarded as essentially criminal. Thus, criminal anthropology provided a powerful argument for racism and imperialism at the height of European colonial expansion. Lombroso, in noting a reduced sensitivity to pain among criminals, wrote:

> Their physical insensibility well recalls that of savage peoples who can bear in rites of puberty, tortures that a

> white man could never endure. All travelers know the indifference of Negroes and American savages to pain: the former cut their hands and laugh in order to avoid work; the latter, tied to the torture post, gaily sing the praises of their tribe while they are slowly burnt. [You can't beat a racist a priori. Think of how many Western heroes died bravely in excruciating pain—Saint Joan burned, Saint Sebastian transfixed with arrows, other martyrs racked, drawn, and quartered. But when an Indian fails to scream and beg for mercy, it can only mean that he doesn't feel the pain.]

If Lombroso and his colleagues had been a dedicated group of proto-Nazis, we could dismiss the whole phenomenon as a ploy of conscious demagogues. It would then convey no other message than a plea for vigilance against ideologues who misuse science. But the leaders of criminal anthropology were "enlightened" socialists and social democrats who viewed their theory as the spearhead for a rational, scientific society based on human realities. The genetic determination of criminal action, Lombroso argued, is simply the law of nature and of evolution:

> We are governed by silent laws which never cease to operate and which rule society with more authority than the laws inscribed on our statute books. Crime appears to be a natural phenomenon . . . like birth or death.

In retrospect, Lombroso's scientific "reality" turned out to be his social prejudice imposed before the fact upon a supposedly objective study. His notions condemned many innocent people to a prejudgment that often worked as a self-fulfilling prophecy. His attempt to understand human behavior by mapping an innate potential displayed in our anatomy served only to work against social reform by placing all blame upon a criminal's inheritance.

Of course, no one takes the claims of Lombroso seriously today. His statistics were faulty beyond belief; only a blind faith in inevitable conclusions could have led to his fudging and finagling. Besides, no one would look to long arms and jutting jaws today as signs of inferiority; modern determinists seek a more fundamental marker in genes and chromosomes.

Much has happened in the 100 years between *L'uomo delinquente* and our Bicentennial celebrations. No serious advocate of innate criminality recommends the irrevocable detention or murder of the unfortunately afflicted or even claims that a natural penchant for criminal behavior necessarily leads to criminal action. Still, the spirit of Lombroso is very much with us. When Richard Speck murdered eight nurses in Chicago, his defense argued that he couldn't help it because he bore an extra Y chromosome. (Normal females have two X chromosomes, normal males an X and a Y. A small percentage of males have an extra Y chromosome, XYY.) This revelation inspired a rash of speculation; articles on the "criminal chromosome" inundated our popular magazines. The naïvely determinist argument had little going for it beyond the following: Males tend to be more aggressive than females; this may be genetic. If genetic, it must reside on the Y chromosome; anyone possessing two Y chromosomes has a double dose of aggressiveness and might incline to violence and criminality. But the hastily collected information on XYY males in prisons seems hopelessly ambiguous, and even Speck himself turns out to be an XY male after all. Once again, biological determinism makes a splash, creates a wave of discussion and cocktail party chatter, and then dissipates for want of evidence. Why are we so intrigued by hypotheses about innate disposition? Why do we wish to fob off responsibility for our violence and sexism upon our genes? The hallmark of humanity is not only our mental capacity but also our mental flexibility. We have made our world and we can change it.

8 The Science and Politics of Human Nature

Part A | Race, Sex and Violence

29 Why We Should Not Name Human Races—A Biological View

TAXONOMY IS THE study of classification. We apply rigorous rules of taxonomy to other forms of life, but when we get to the species we should know best, we have particular problems.

We commonly divide our own species into races. Under the rules of taxonomy, all formal subdivisions of species are called subspecies. Human races, therefore, are subspecies of *Homo sapiens.*

During the past decade, the practice of dividing species into subspecies has been gradually abandoned in many quarters, as the introduction of quantitative techniques suggests different methods for the study of geographic variability within species. The designation of human races cannot and should not be divorced from social and ethical questions pertaining to our species alone. Nonetheless, these new taxonomic procedures add a general and purely biological argument to an old debate.

I contend that the continued racial classification of *Homo sapiens* represents an outmoded approach to the general problem of differentiation within a species. In other words, I reject a racial classification of humans for the same reasons that I prefer not to divide into subspecies the prodigiously variable West Indian land snails that form the subject of my own research.

The argument against racial classification has been made many times before, notably by eleven authors in *The Concept of Race,* a book edited by Ashley Montagu in 1964 (republished in 1969 as a Collier-Macmillan paperback). Yet these views did not command general assent because taxonomic practice a decade ago still supported the routine designation of subspecies. In 1962, for example, Theodosius Dobzhansky expressed astonishment that "some authors have talked themselves into denying that the human species has any races at all! . . . Just as zoologists observe a great diversity of animals, anthropologists are confronted with a diversity of human beings. . . . Race is the subject of scientific study and analysis simply because it is a fact of nature." And Grant Bogue, in a debate with Ashley Montagu, recently wrote: "Some misfit academicians have said no, that this is all a mistake . . . and some have gone so far as to suggest that even the very concept of race is all in our heads. . . . To this contention there are several answers. One is often voiced: race is self-evident."

There is a glaring fallacy in these arguments. Geographic variability, not race, is self-evident. No one can deny that *Homo sapiens* is a strongly differentiated species; few will quarrel with the observation that differences in skin color are the most striking outward sign of this variability. But the fact of variability does not require the designation of races. There are better ways to study human differences.

The category of species has a special status in the taxonomic hierarchy. Under tenets of the "biological species concept," each species represents a "real" unit in nature. Its definition reflects this status: "a population of actually or potentially interbreeding organisms sharing a common gene pool." Above the species level, we encounter a certain arbitrariness. One man's genus may be another man's family. Nonetheless, there are certain rules that must be followed in the construction of hierarchies. You cannot, for example, place two members of the same taxon (genus, for example) into different taxa of a still higher category (family or order, for example).

Below the species level, we have only the subspecies. In *Systematics and the Origin of Species* (Columbia University Press, 1942), Ernst Mayr defined this category: "The subspecies, or geographic race, is a geographically localized subdivision of the species, which differs genetically and taxonomically from other subdivisions of the species." We need to satisfy two criteria: (1) A subspecies must be recognizable by features of its morphology, physiology, or behavior, that is, it must be "taxonomically" (and by inference, genetically) different from other subspecies; and (2) A subspecies must occupy a subdivision of the total geographic range of the species. When we decide to characterize variation within a species by establishing subspecies, we partition a spectrum of variation into discrete packages with distinct geographic borders and recognizable traits.

The subspecies differs from all other taxonomic categories in two fundamental ways: (1) Its boundaries can never be fixed and definite because, by definition, a member of one subspecies can interbreed with members of any other subspecies in its species (a group that cannot breed with other closely related forms must be designated as a full species); (2) The category need not be used. All organisms must belong to a species, each species must belong to a genus, each genus to a family, and so on. But there is no requirement that a species be divided into subspecies. The subspecies is a category of convenience. We use it only when we judge that our understanding of variability will be increased by establishing discrete, geographically bounded packages within a species. Many biologists are now arguing that it is not only inconvenient, but also downright misleading, to impose a formal nomenclature on the dynamic patterns of variability that we observe in nature.

How can we deal with the rich geographic variability that characterizes so many species, including our own? As an example of the old approach, a monograph was published in 1942 on geographic variation in the Hawaiian tree snail *Achatinella apexfulva.* The author divided this astonishingly variable species into seventy-eight formal subspecies and sixty additional "microgeographic races" (for units a bit too indistinct for subspecific status). Each subdivision received a name and a formal description. The result is a voluminous and almost unreadable tome that buries one of the most interesting phenomena of evolutionary biology under an impenetrable thicket of names and static descriptions.

And yet there are patterns of variation within this species that would fascinate any biologist: correlations of shell form with altitude and rainfall, variation subtly attuned to climatic conditions, routes of migration reflected in the distribution of color markings on the shell. Shall our approach to such variation be that of a cataloger? Shall we artificially partition such a dynamic and continuous pattern into distinct units with formal names? Would it not be better to map this variability objectively without imposing upon it the subjective criteria for formal subdivision that any taxonomist must use in naming subspecies?

I think that most biologists would now answer "yes" to my last question; I also think that they would have given the same answer thirty years ago. Why, then, did they continue to treat geographic variation by establishing subspecies? They did so because objective techniques had not been developed for mapping the continuous variation of a species. They could, to be sure, map the distribution of single characters, for example, body weight. But variation in single traits is a pale shadow of patterns in variation that affect so many features simultaneously. Moreover, the classical problem of "incongruity" arises. Maps constructed for other single traits almost invariably present different distributions: size may be large in cold climates and small in warm, while color may be light in open country and dark in forests.

A satisfactory procedure for objective mapping demands that variation in many characters be treated simultaneously. This simultaneous treatment is called "multivariate analysis." Statisticians developed the basic theories of multivariate analysis many years ago, but its routine use could not even be contemplated before the invention of large electronic computers. The computations involved are extremely laborious and quite beyond the capacities of desk calculators and human patience; but the computer can perform them in seconds.

During the past decade, studies of geographic variation have been transformed by the use of multivariate analysis. Almost all the proponents of multivariate analysis have declined to name subspecies. You cannot map a continuous distribution if all specimens must first be allocated to discrete subdivisions. Is it not better simply to characterize each local sample by its own morphology and to search for interesting regularities in the maps so produced?

The English sparrow, for example, was introduced into

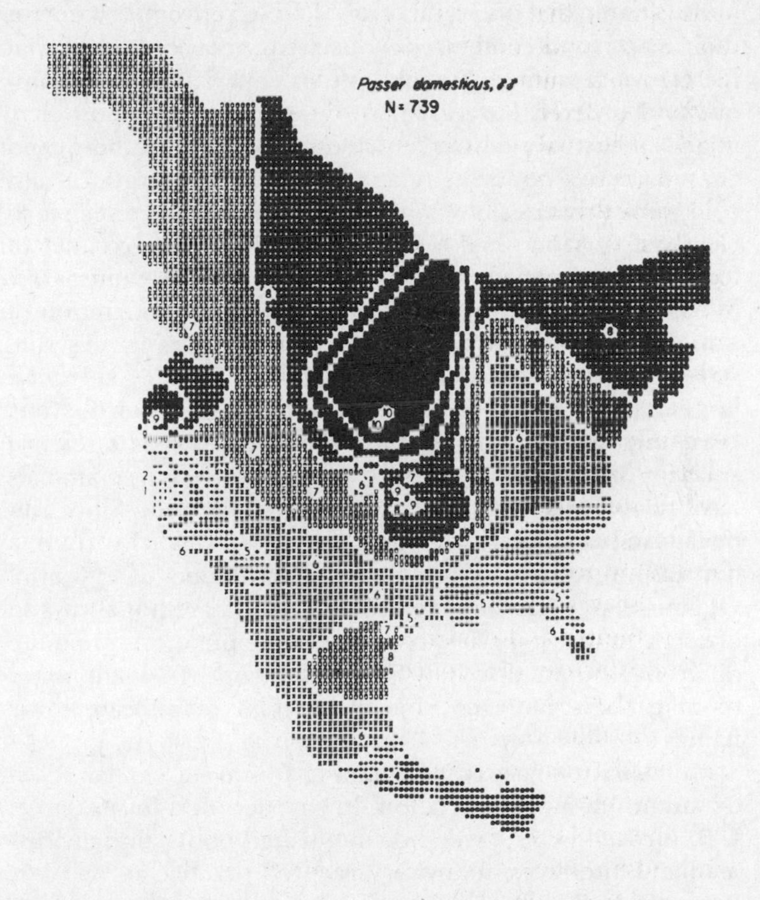

A map generated and drawn by a computer shows the distribution by size of male house sparrows in North America. The higher numbers indicate larger sizes, based on a composite of sixteen different measurements of the birds' skeletons.

North America in the 1850s. Since then it has spread geographically and differentiated morphologically to a remarkable degree. Previously, this variation was treated by naming subspecies. R. F. Johnston and R. K. Selander (in *Science*, 1964, p. 550) refused to follow this procedure. "We are not convinced," they argued, "that nomenclatural stasis is desirable for a patently dynamic system." Instead, they mapped multivariate patterns of variation. I have reproduced one of their maps for a combination of sixteen morphological characters representing general body size. Variation is continuous and orderly. Large sparrows tend to live in northern, inland areas, while small sparrows inhabit southern and coastal areas. The strong relationship between large size and cold winter climates is obvious. But would we have seen it so clearly if variation had been expressed instead by a set of formal Latin names artificially dividing the continuum? Moreover, this pattern of variation reflects the operation of a major principle of animal distribution. Bergmann's rule states that members of a warm-blooded species tend to be larger in cold climates. The standard explanation for this regularity invokes the relationship between size and relative surface (discussed in the essays of section 6). Large animals have relatively less surface area than smaller ones. Since animals lose heat by radiation through their external surface, a decrease in relative surface area helps the body keep warm. Of course, patterns of geographic variation are not always so orderly. In many species, certain local populations are quite different from immediately adjacent groups. It is still better to map these patterns objectively than to allocate static names.

Multivariate analysis is beginning to have a similar effect on studies of human variation. In past decades, for example, J. B. Birdsell wrote several distinguished books that divided mankind into races, following accepted practice at the time. Recently he has applied multivariate analysis to gene frequencies for blood types among Australian Aborigines. He rejects any subdivision into discrete units and writes: "It may well be that the investigation of the nature and intensity of evolutionary forces is the endeavor to be pursued while the pleasures of classifying man fall away, perhaps forever."

30 The Nonscience of Human Nature

WHEN A GROUP of girls suffered simultaneous seizures in the presence of an accused witch, the justices of seventeenth century Salem could offer no explanation other then true demonic possession. When the followers of Charlie Manson attributed occult powers to their leader, no judge took them seriously. In nearly three hundred years separating the two incidents, we have learned quite a bit about social, economic, and psychological determinants of group behavior. A crudely literal interpretation of such events now seems ridiculous.

An equally crude literalism used to prevail in interpreting human nature and the differences among human groups. Human behavior was attributed to innate biology; we do what we do because we are made that way. The first lesson of an eighteenth-century primer stated the position succinctly: In Adam's fall, we sinned all. A movement away from this biological determinism has been a major trend in twentieth-century science and culture. We have come to see ourselves as a learning animal; we have come to believe that the influences of class and culture far outweigh the weaker predispositions of our genetic constitution.

Nonetheless, we have been deluged during the past decade by a resurgent biological determinism, ranging from "pop ethology" to outright racism.

With Konrad Lorenz as godfather, Robert Ardrey as dramatist, and Desmond Morris as raconteur, we are presented with man, "the naked ape," descended from an African carnivore, innately aggressive and inherently territorial.

Lionel Tiger and Robin Fox try to find a biological basis for outmoded Western ideals of aggressive, outreaching men and docile, restricted women. In discussing cross-cultural differences between men and women, they propose a hormonal chemistry inherited from the requirements of our supposed primal roles as group hunters and child rearers.

Carleton Coon offered a prelude of events to come with his claim (*The Origin of Races,* 1962) that five major human races evolved independently from *Homo erectus* ("Java" and "Peking" man) to *Homo sapiens,* with black people making the transition last. More recently, the IQ test has been (mis)used to infer genetic differences in intelligence among races (Arthur Jensen and William Shockley) and classes (Richard Herrnstein)—always, I must note, to the benefit of the particular group to which the author happens to belong (see next essay).

All these views have been ably criticized on an individual basis; yet they have rarely been treated together as expressions of a common philosophy—a crude biological determinism. One can, of course, accept a specific claim and reject the others. A belief in the innate nature of human violence does not brand anyone a racist. Yet all these claims have a common underpinning in postulating a direct genetic basis for our most fundamental traits. If we are programmed to be what we are, then these traits are ineluctable. We may, at best, channel them, but we cannot change them, either by will, education, or culture.

If we accept the usual platitudes about "scientific method" at face value, then the coordinated resurgence of biological determinism must be attributed to new information that refutes the earlier findings of twentieth-century science. Science, we are told, progresses by accumulating new information and using it to improve or replace old theories. But the new biological determinism rests upon no recent fund of information and can cite in its behalf not a single unambiguous fact. Its renewed support must have some other basis, most likely social or political in nature.

Science is always influenced by society, but it operates

under a strong constraint of fact as well. The Church eventually made its peace with Galileo because, after all, the earth does go around the sun. In studying the genetic components of such complex human traits as intelligence and aggressiveness, however, we are freed from the constraint of fact, for we know practically nothing. In these questions, "science" follows (and exposes) the social and political influences acting upon it.

What then, are the nonscientific reasons that have fostered the resurgence of biological determinism? They range, I believe, from pedestrian pursuits of high royalties for best sellers to pernicious attempts to reintroduce racism as respectable science. Their common denominator must lie in our current malaise. How satisfying it is to fob off the responsibility for war and violence upon our presumably carnivorous ancestors. How convenient to blame the poor and the hungry for their own condition—lest we be forced to blame our economic system or our government for an abject failure to secure a decent life for all people. And how convenient an argument for those who control government and, by the way, provide the money that science requires for its very existence.

Deterministic arguments divide neatly into two groups—those based on the supposed nature of our species in general and those that invoke presumed differences among "racial groups" of *Homo sapiens.* I discuss the first subject here and treat the second in my next essay.

Summarized briefly, mainstream pop ethology contends that two lineages of hominids inhabited Pleistocene Africa. One, a small, territorial carnivore, evolved into us; the other, a larger, presumably gentle herbivore, became extinct. Some carry the analogy of Cain and Abel to its full conclusion and accuse our ancestors of fratricide. The "predatory transition" to hunting established a pattern of innate violence and engendered our territorial urges: "With the coming of the hunting life to the emerging hominid came the dedication to territory" (Ardrey, *The Territorial Imperative*). We may be clothed, citified, and civilized, but we carry deep within us the genetic patterns of behavior that served our ancestor, the "killer ape." In *African Genesis* Ardrey champions Raymond Dart's contention that "the predatory transition and the weapons fixation explained man's bloody history, his eternal aggression, his irrational, self-destroying, inexorable pursuit of death for death's sake."

Tiger and Fox extend the theme of group hunting to proclaim a biological basis for the differences between men and women that Western cultures have traditionally valued. Men did the hunting; women stayed home with the kids. Men are aggressive and combative, but they also form strong bonds among themselves that reflect the ancient need for cooperation in the killing of big game and now find expression in touch football and rotary clubs. Women are docile and devoted to their own children. They do not form intense bonds among themselves because their ancestors needed none to tend their homes and their men: sisterhood is an illusion. "We are wired for hunting. . . . We remain Upper Paleolithic hunters, fine-honed machines designed for the efficient pursuit of game" (Tiger and Fox, *The Imperial Animal*).

The story of pop ethology has been built on two lines of supposed evidence, both highly disputable:

1. Analogies with the behavior of other animals (abundant but imperfect data). No one doubts that many animals (including some, but not all, primates) display innate patterns of aggression and territorial behavior. Since we exhibit similar behavior, can we not infer a similar cause? The fallacy of this assumption reflects a basic issue in evolutionary theory. Evolutionists divide the similarities between two species into *homologous* features shared by common descent and a common genetic constitution, and *analogous* traits evolved separately.

Comparisons between humans and other animals lead to causal assertions about the genetics of our behavior only if they are based on homologous traits. But how can we know whether similarities are homologous or analogous? It is hard to differentiate even when we deal with concrete structures, such as muscles and bones. In fact, most classical arguments in the study of phylogeny involve the confusion of homology and analogy, for analogous structures can be strikingly similar (we call this phenomenon evolutionary convergence). How much harder it is to tell when similar features are only the outward motions of behavior! Baboons may be territorial; their males may be organized into a dominance hierarchy—but is our quest for Lebensraum and the hierarchy of our armies an expression of the same genetic makeup or merely an analogous pattern that might be purely cultural in origin? And when Lorenz compares us with geese and fish, we stray even further into pure conjecture; baboons, at least, are second cousins.

2. Evidence from hominid fossils (scrappy but direct data). Ardrey's claims for territoriality rest upon the assumption that our African ancestor *Australopithecus africanus,* was a carnivore. He derives his "evidence" from accumulations of bones and tools at the South African cave sites and the size and shape of teeth. The bone piles are no longer seriously considered; they are more likely the work of hyenas than of hominids.

Teeth are granted more prominence, but I believe that the evidence is equally poor if not absolutely contradictory. The argument rests upon relative size of grinding teeth (premolars and molars). Herbivores need more surface area to grind their gritty and abundant food. *A. robustus,* the supposed gentle herbivore, possessed grinding teeth relatively larger than those of its carnivorous relative, our ancestor *A. africanus.*

But *A. robustus* was a larger creature than *A. africanus.* As size increases, an animal must feed a body growing as the cube of length by chewing with tooth areas that increase only as the square of length if they maintain the same relative size (see essays of section 6). This will not do, and larger mammals must have differentially larger teeth than smaller relatives. I have tested this assertion by measuring tooth areas and body sizes for species in several groups of mammals (rodents, piglike herbivores, deer, and several groups of primates). Invariably, I find that larger animals have relatively larger teeth—not because they eat different foods, but simply because they are larger.

Moreover, the "small" teeth of *A. africanus* are not at all diminutive. They are *absolutely larger* than ours (although we are three times as heavy), and they are about as big as those of gorillas weighing nearly ten times as much! The evidence of tooth size indicates to me that *A. africanus* was primarily herbivorous.

The issue of biological determinism is not an abstract matter to be debated within academic cloisters. These ideas have important consequences, and they have already permeated our mass media. Ardrey's dubious theory is a prominent theme in Stanley Kubrick's film *2001.* The bone tool of our apelike ancestor first smashes a tapir's skull and then twirls about to transform into a space station of our next evolutionary stage—as the superman theme of Richard Strauss' *Zarathustra* yields to Johann's *Blue Danube.* Kubrick's next film, *Clockwork Orange,* continues the theme and explores the dilemma inspired by claims of innate human violence. (Shall we accept totalitarian controls for mass deprogramming or remain nasty and vicious within a democracy?) But the most immediate impact will be felt as male privilege girds its loins to battle a growing women's movement. As Kate Millett remarks in *Sexual Politics:* "Patriarchy has a tenacious or powerful hold through its successful habit of passing itself off as nature."

31 | Racist Arguments and IQ

LOUIS AGASSIZ, the greatest biologist of mid-nineteenth-century America, argued that God had created blacks and whites as separate species. The defenders of slavery took much comfort from this assertion, for biblical prescriptions of charity and equality did not have to extend across a species boundary. What could an abolitionist say? Science had shone its cold and dispassionate light upon the subject; Christian hope and sentimentality could not refute it.

Similar arguments, carrying the apparent sanction of science, have been continually invoked in attempts to equate egalitarianism with sentimental hope and emotional blindness. People who are unaware of this historical pattern tend to accept each recurrence at face value: that is, they assume that each statement arises from the "data" actually presented, rather than from the social conditions that truly inspire it.

The racist arguments of the nineteenth century were based primarily on craniometry, the measurement of human skulls. Today, these contentions stand totally discredited. What craniometry was to the nineteenth century, intelligence testing has been to the twentieth. The victory of the eugenics movement in the Immigration Restriction Act of 1924 signaled its first unfortunate effect—for the severe restrictions upon non-Europeans and upon southern and eastern Europeans gained much support from results of the first extensive and uniform application of intelligence tests in America—the Army Mental Tests of World War I. These tests were engineered and administered by psychologist Robert M. Yerkes, who concluded that "education alone will not place the negro [*sic*] race on a par with its Caucasian competitors." It is now clear that Yerkes and his colleagues knew no way to separate genetic from environmental components in postulating causes for different performances on the tests.

The latest episode of this recurring drama began in 1969, when Arthur Jensen published an article entitled, "How Much Can We Boost IQ and Scholastic Achievement?" in the *Harvard Educational Review.* Again, the claim went forward that new and uncomfortable information had come to light, and that science had to speak the "truth" even if it refuted some cherished notions of a liberal philosophy. But again, I shall argue, Jensen had no new data; and what he did present was flawed beyond repair by inconsistencies and illogical claims.

Jensen assumes that IQ tests adequately measure something we may call "intelligence." He then attempts to tease apart the genetic and environmental factors causing differences in performance. He does this primarily by relying upon the one natural experiment we possess: identical twins reared apart—for differences in IQ between genetically identical people can only be environmental. The average difference in IQ for identical twins is less than the difference for two unrelated individuals raised in similarly varied environments. From the data on twins, Jensen obtains an estimate of environmental influence. He concludes that IQ has a heritability of about 0.8 (or 80 percent) *within* the population of American and European whites. The average difference between American whites and blacks is 15 IQ points (one standard deviation). He asserts that this difference is too large to attribute to environment, given the high heritability of IQ. Lest anyone think that Jensen writes in the tradition of abstract scholarship, I merely quote the first line of his famous work: "Compensatory education has been tried, and it apparently has failed."

I believe that this argument can be refuted in a "hierarchical" fashion—that is, we can discredit it at one level and then show that it fails at a more inclusive level even if we allow Jensen's argument for the first two levels:

Level 1: The equation of IQ with intelligence. Who knows what IQ measures? It is a good predictor of "success" in school, but is such success a result of intelligence, apple polishing, or the assimilation of values that the leaders of society prefer? Some psychologists get around this argument by defining intelligence operationally as the scores attained on "intelligence" tests. A neat trick. But at this point, the technical definition of intelligence has strayed so far from the vernacular that we can no longer define the issue. But let me allow (although I don't believe it), for the sake of argument, that IQ measures some meaningful aspect of intelligence in its vernacular sense.

Level 2: The heritability of IQ. Here again, we encounter a confusion between vernacular and technical meanings of the same word. "Inherited," to a layman, means "fixed," "inexorable," or "unchangeable." To a geneticist, "inherited" refers to an estimate of similarity between related individuals based on genes held in common. It carries no implications of inevitability or of immutable entities beyond the reach of environmental influence. Eyeglasses correct a variety of inherited problems in vision; insulin can check diabetes.

Jensen insists that IQ is 80 percent heritable. Princeton psychologist Leon J. Kamin has done the dog-work of meticulously checking through details of the twin studies that form the basis of this estimate. He has found an astonishing number of inconsistencies and downright inaccuracies. For example, the late Sir Cyril Burt, who generated the largest body of data on identical twins reared apart, pursued his studies of intelligence for more than forty years. Although he increased his sample sizes in a variety of "improved" versions, some of his correlation coefficients remain unchanged to the third decimal place—a statistically impossible situation.[5] IQ depends in part upon sex and age; and other studies

---

5 | I wrote this essay in 1974. Since then, the case against Sir Cyril has progressed from an inference of carelessness to a spectacular

did not standardize properly for them. An improper correction may produce higher values between twins not because they hold genes for intelligence in common, but simply because they share the same sex and age. The data are so flawed that no valid estimate for the heritability of IQ can be drawn at all. But let me assume (although no data support it), for the sake of argument, that the heritability of IQ is as high as 0.8.

Level 3 : The confusion of within- and between-group variation. Jensen draws a causal connection between his two major assertions—that the within-group heritability of IQ is 0.8 for American whites, and that the mean difference in IQ between American blacks and whites is 15 points. He assumes that the black "deficit" is largely genetic in origin because IQ is so highly heritable. This is a *non sequitur* of the worst possible kind—for there is no necessary relationship between heritability within a group and differences in mean values of two separate groups.

A simple example will suffice to illustrate this flaw in Jensen's argument. Height has a much higher heritability within groups than anyone has ever claimed for IQ. Suppose that height has a mean value of five feet two inches and a heritability of 0.9 (a realistic value) within a group of nutritionally deprived Indian farmers. High heritability simply means that short farmers will tend to have short offspring, and tall farmers tall offspring. It says nothing whatever against the possibility that proper nutrition could raise the mean height to six feet (taller than average white Americans). It only means that, in this improved status, farmers shorter than average (they may now be five feet ten inches) would still tend to have shorter than average children.

I do not claim that intelligence, however defined, has no genetic basis—I regard it as trivially true, uninteresting, and

---

(and well-founded) suspicion of fraud. Reporters for the London *Times* have discovered, for example, that Sir Cyril's coauthors (for the infamous twin studies) apparently did not exist outside his imagination. In the light of Kamin's discoveries, one must suspect that the data have an equal claim to reality.

unimportant that it does. The expression of any trait represents a complex interaction of heredity and environment. Our job is simply to provide the best environmental situation for the realization of valued potential in all individuals. I merely point out that a specific claim purporting to demonstrate a mean genetic deficiency in the intelligence of American blacks rests upon no new facts whatever and can cite no valid data in its support. It is just as likely that blacks have a genetic advantage over whites. And, either way, it doesn't matter a damn. An individual can't be judged by his group mean.

If current biological determinism in the study of human intelligence rests upon no new facts (actually, no facts at all), then why has it become so popular of late? The answer must be social and political. The 1960s were good years for liberalism; a fair amount of money was spent on poverty programs and relatively little happened. Enter new leaders and new priorities. Why didn't the earlier programs work? Two possibilities are open: (1) we didn't spend enough money, we didn't make sufficiently creative efforts, or (and this makes any established leader jittery) we cannot solve these problems without a fundamental social and economic transformation of society; or (2) the programs failed because their recipients are inherently what they are—blaming the victims. Now, which alternative will be chosen by men in power in an age of retrenchment?

I have shown, I hope, that biological determinism is not simply an amusing matter for clever cocktail party comments about the human animal. It is a general notion with important philosophical implications and major political consequences. As John Stuart Mill wrote, in a statement that should be the motto of the opposition: "Of all the vulgar modes of escaping from the consideration of the effect of social and moral influences upon the human mind, the most vulgar is that of attributing the diversities of conduct and character to inherent natural differences."

Part B | Sociobiology

32 Biological Potentiality vs. Biological Determinism

IN 1758, LINNAEUS faced the difficult decision of how to classify his own species in the definitive edition of his *Systema Naturae.* Would he simply rank *Homo sapiens* among the other animals or would he create for us a separate status? Linnaeus compromised. He placed us within his classification (close to monkeys and bats), but set us apart by his description. He defined our relatives by the mundane, distinguishing characters of size, shape, and number of fingers and toes. For *Homo sapiens,* he wrote only the Socratic injunction: *nosce te ipsum*—"know thyself."

For Linnaeus, *Homo sapiens* was both special and not special. Unfortunately, this eminently sensible resolution has been polarized and utterly distorted by most later commentators. Special and not special have come to mean nonbiological and biological, or nurture and nature. These later polarizations are nonsensical. Humans are animals and everything we do lies within our biological potential. Nothing arouses this ardent (although currently displaced) New Yorker to greater anger than the claims of some self-styled "ecoactivists" that large cities are the "unnatural" harbingers of our impending destruction. But—and here comes the biggest *but* I can muster—the statement that humans are animals does not imply that our specific patterns of behavior and social arrangements are in any way directly determined by our genes. *Potentiality* and *determination* are different concepts.

The intense discussion aroused by E. O. Wilson's *Sociobi-*

ology (Harvard University Press, 1975) has led me to take up this subject. Wilson's book has been greeted by a chorus of praise and publicity. I, however, find myself among the smaller group of its detractors. Most of *Sociobiology* wins from me the same high praise almost universally accorded it. For a lucid account of evolutionary principles and an indefatigably thorough discussion of social behavior among all groups of animals, *Sociobiology* will be the primary document for years to come. But Wilson's last chapter, "From Sociobiology to Sociology," leaves me very unhappy indeed. After twenty-six chapters of careful documentation for the nonhuman animals, Wilson concludes with an extended speculation on the genetic basis of supposedly universal patterns in human behavior. Unfortunately, since this chapter is his statement on the subject closest to all our hearts, it has also attracted more than 80 percent of all the commentary in the popular press.

We who have criticized this last chapter have been accused of denying altogether the relevance of biology to human behavior, of reviving an ancient superstition by placing ourselves outside the rest of "the creation." Are we pure "nurturists?" Do we permit a political vision of human perfectibility to blind us to evident constraints imposed by our biological nature? The answer to both statements is no. The issue is not universal biology vs. human uniqueness, but biological potentiality vs. biological determinism.

Replying to a critic of his article in the *New York Times Magazine* (October 12, 1975), Wilson wrote:

> There is no doubt that the patterns of human social behavior, including altruistic behavior, are under genetic control, in the sense that they represent a restricted subset of possible patterns that are very different from the patterns of termites, chimpanzees and other animal species.

If this is all that Wilson means by genetic control, then we can scarcely disagree. Surely we do not do all the things that other animals do, and just as surely, the range of our potential behavior is circumscribed by our biology. We would lead very different social lives if we photosynthesized (no agriculture, gathering, or hunting—the major determinants of our social evolution) or had life cycles like those of the gall midges discussed in essay 10. (When feeding on an uncrowded mushroom, these insects reproduce in the larval or pupal stage. The young grow within the mother's body, devour her from inside, and emerge from her depleted external shell ready to feed, grow the next generation, and make the supreme sacrifice.)

But Wilson makes much stronger claims. Chapter 27 is not a statement about the range of potential human behaviors or even an argument for the restriction of that range from a much larger total domain among all animals. It is, primarily, an extended speculation on the existence of genes for specific and variable traits in human behavior—including spite, aggression, xenophobia, conformity, homosexuality, and the characteristic behavioral differences between men and women in Western society. Of course, Wilson does not deny the role of nongenetic learning in human behavior; he even states at one point that "genes have given away most of their sovereignty." But, he quickly adds, genes "maintain a certain amount of influence in at least the behavioral qualities that underlie variations between cultures." And the next paragraph calls for a "discipline of anthropological genetics."

Biological determinism is the primary theme in Wilson's discussion of human behavior; chapter 27 makes no sense in any other context. Wilson's primary aim, as I read him, is to suggest that Darwinian theory might reformulate the human sciences just as it previously transformed so many other biological disciplines. But Darwinian processes can not operate without genes to select. Unless the "interesting" properties of human behavior are under specific genetic control, sociology need fear no invasion of its turf. By interesting, I refer to the subjects sociologists and anthropologists fight about most often—aggression, social stratification, and differences in behavior between men and women. If genes only specify that we are large enough to live in a world of gravitational forces, need to rest our bodies by sleeping, and do not photosynthesize, then the realm of genetic determinism will be relatively uninspiring.

What is the direct evidence for genetic control of specific human social behavior? At the moment, the answer is none whatever. (It would not be impossible, in theory, to gain such evidence by standard, controlled experiments in breeding, but we do not raise people in *Drosophila* bottles, establish pure lines, or control environments for invariant nurturing.) Sociobiologists must therefore advance indirect arguments based on plausibility. Wilson uses three major strategies: universality, continuity, and adaptiveness.

1. Universality: If certain behaviors are invariably found in our closest primate relatives and among humans themselves, a circumstantial case for common, inherited genetic control may be advanced. Chapter 27 abounds with statements about supposed human universals. For example, "Human beings are absurdly easy to indoctrinate—they *seek* it." Or, "Men would rather believe than know." I can only say that my own experience does not correspond with Wilson's.

When Wilson must acknowledge diversity, he often dismisses the uncomfortable "exceptions" as temporary and unimportant aberrations. Since Wilson believes that repeated, often genocidal warfare has shaped our genetic destiny, the existence of nonaggressive peoples is embarrassing. But he writes: "It is to be expected that some isolated cultures will escape the process for generations at a time, in effect reverting temporarily to what ethnographers classify as a pacific state."

In any case, even if we can compile a list of behavioral traits shared by humans and our closest primate relatives, this does not make a good case for common genetic control. Similar results need not imply similar causes; in fact, evolutionists are so keenly aware of this problem that they have developed a terminology to express it. Similar features due to common genetic ancestry are "homologous"; similarities due to common function, but with different evolutionary histories, are "analogous" (the wings of birds and insects, for example—the common ancestor of both groups lacked wings). I will argue below that a basic feature of human biology supports the idea that many behavioral similarities between humans and other primates are analogous, and that they have no direct genetic specification in humans.

2. Continuity: Wilson claims, with ample justice in my opinion, that the Darwinian explanation of altruism in W. D. Hamilton's 1964 theory of "kin selection" forms the basis for an evolutionary theory of animal societies. Altruistic acts are the cement of stable societies, yet they seem to defy a Darwinian explanation. On Darwinian principles, all individuals are selected to maximize their own genetic contribution to future generations. How, then, can they willingly sacrifice or endanger themselves by performing altruistic acts to benefit others?

The resolution is charmingly simple in concept, although complex in technical detail. By benefiting relatives, altruistic acts preserve an altruist's genes even if the altruist himself will not be the one to perpetuate them. For example, in most sexually reproducing organisms, an individual shares (on average) one-half the genes of his sibs and one-eighth the genes of his first cousins. Hence, if faced with a choice of saving oneself alone or sacrificing oneself to save more than two sibs or more than eight first cousins, the Darwinian calculus favors altruistic sacrifice; for in so doing, an altruist actually increases his own genetic representation in future generations.

Natural selection will favor the preservation of such self-serving altruist genes. But what of altruistic acts toward non-relatives? Here sociobiologists must invoke a related concept of "reciprocal altruism" to preserve a genetic explanation. The altruistic act entails some danger and no immediate benefit, but if it inspires a reciprocal act by the current beneficiary at some future time, it may pay off in the long run: a genetic incarnation of the age-old adage: you scratch my back and I'll scratch yours (even if we're not related).

The argument from continuity then proceeds. Altruistic acts in other animal societies can be plausibly explained as examples of Darwinian kin selection. Humans perform altruistic acts and these are likely to have a similarly direct genetic basis. But again, similarity of result does not imply identity of cause (see below for an alternate explanation based on biological potentiality rather than biological determinism).

3. Adaptiveness: Adaptation is the hallmark of Darwinian processes. Natural selection operates continuously and re-

lentlessly to fit organisms to their environments. Disadvantageous social structures, like poorly designed morphological structures, will not survive for long.

Human social practices are clearly adaptive. Marvin Harris has delighted in demonstrating the logic and sensibility of those social practices in other cultures that seem most bizarre to smug Westerners (*Cows, Pigs, Wars, and Witches.* Random House, 1974). Human social behavior is riddled with altruism; it is also clearly adaptive. Is this not a prima facie argument for direct genetic control? My answer is definitely "no," and I can best illustrate my claim by reporting an argument I recently had with an eminent anthropologist.

My colleague insisted that the classic story of Eskimos on ice floes provides adequate proof for the existence of specific altruist genes maintained by kin selection. Apparently, among some Eskimo peoples, social units are arranged as family groups. If food resources dwindle and the family must move to survive, aged grandparents willingly remain behind (to die) rather than endanger the survival of their entire family by slowing an arduous and dangerous migration. Family groups with no altruist genes have succumbed to natural selection as migrations hindered by the old and sick lead to the death of entire families. Grandparents with altruist genes increase their own fitness by their sacrifice, for they enhance the survival of close relatives sharing their genes.

The explanation by my colleague is plausible, to be sure, but scarcely conclusive since an eminently simple, nongenetic explanation also exists: there are no altruist genes at all, in fact, no important genetic differences among Eskimo families whatsoever. The sacrifice of grandparents is an adaptive, but nongenetic, cultural trait. Families with no tradition for sacrifice do not survive for many generations. In other families, sacrifice is celebrated in song and story; aged grandparents who stay behind become the greatest heroes of the clan. Children are socialized from their earliest memories to the glory and honor of such sacrifice.

I cannot prove my scenario, any more than my colleague can demonstrate his. But in the current context of no evidence, they are at least equally plausible. Likewise, reciprocal altruism undeniably exists in human societies, but this provides no evidence whatever for its genetic basis. As Benjamin Franklin said: "We must all hang together, or assuredly we shall all hang separately." Functioning societies may require reciprocal altruism. But these acts need not be coded into our consciousness by genes; they may be inculcated equally well by learning.

I return, then, to Linnaeus's compromise—we are both ordinary and special. The central feature of our biological uniqueness also provides the major reason for doubting that our behaviors are directly coded by specific genes. That feature is, of course, our large brain. Size itself is a major determinant of the function and structure of any object. The large and the small cannot work in the same way (see section 6). The study of changes that accompany increasing size is called "allometry." Best known are the structural changes that compensate for decreasing surface/volume ratios of large creatures—relatively thick legs and convoluted internal surfaces (lungs, and villi of the small intestine, for example). But markedly increased brain size in human evolution may have had the most profound allometric consequences of all—for it added enough neural connections to convert an inflexible and rather rigidly programmed device into a labile organ, endowed with sufficient logic and memory to substitute nonprogrammed learning for direct specification as the ground of social behavior. Flexibility may well be the most important determinant of human consciousness (see essay 7); the direct programming of behavior has probably become inadaptive.

Why imagine that specific genes for aggression, dominance, or spite have any importance when we know that the brain's enormous flexibility permits us to be aggressive or peaceful, dominant or submissive, spiteful or generous? Violence, sexism, and general nastiness *are* biological since they represent one subset of a possible range of behaviors. But peacefulness, equality, and kindness are just as biological—and we may see their influence increase if we can create social structures that permit them to flourish. Thus, my criticism of Wilson does not invoke a nonbiological "environmentalism"; it merely pits the concept of biological potentiality—a brain capable of the full range of human behaviors and rigidly predisposed toward none—against the idea of biological determinism—specific genes for specific behavioral traits.

But why is this academic issue so delicate and explosive? There is no hard evidence for either position, and what difference does it make, for example, whether we conform because conformer genes have been selected or because our general genetic makeup permits conformity as one strategy among many?

The protracted and intense debate surrounding biological determinism has arisen as a function of its social and political message. As I argue in the preceding set of essays, biological determinism has always been used to defend existing social arrangements as biologically inevitable—from "for ye have the poor always with you" to nineteenth-century imperialism to modern sexism. Why else would a set of ideas so devoid of factual support gain such a consistently good press from established media throughout the centuries? This usage is quite out of the control of individual scientists who propose deterministic theories for a host of reasons, often benevolent.

I make no attribution of motive in Wilson's or anyone else's case. Neither do I reject determinism because I dislike its political usage. Scientific truth, as we understand it, must be our primary criterion. We live with several unpleasant biological truths, death being the most undeniable and ineluctable. If genetic determinism is true, we will learn to live with it as well. But I reiterate my statement that no evidence exists to support it, that the crude versions of past centuries have been conclusively disproved, and that its continued popularity is a function of social prejudice among those who benefit most from the status quo.

But let us not saddle *Sociobiology* with the sins of past determinists. What have been its direct results in the first flush of its excellent publicity? At best, we see the beginnings of a line of social research that promises only absurdity by its refusal to consider immediate nongenetic factors. The January 30, 1976, issue of *Science* (America's leading technical journal for scientists) contains an article on panhandling that I would have accepted as satire if it had appeared verbatim in the

National Lampoon. The authors dispatched "panhandlers" to request dimes from various "targets." Results are discussed only in the context of kin selection, reciprocal altruism, and the food-sharing habits of chimps and baboons—nothing on current urban realities in America. As one major conclusion, they find that male panhandlers are "far more successful approaching a single female or a pair of females than a male and female together; they were particularly unsuccessful when approaching a single male or two males together." But not a word about urban fear or the politics of sex—just some statements about chimps and the genetics of altruism (although they finally admit that reciprocal altruism probably does not apply—after all, they argue, what future benefit can one expect from a panhandler).

In the first negative comment on *Sociobiology,* economist Paul Samuelson (*Newsweek,* July 7, 1975) urged sociobiologists to tread softly in the zones of race and sex. I see no evidence that his advice is being heeded. In his *New York Times Magazine* article of October 12, 1975, Wilson writes:

> In hunter-gatherer societies, men hunt and women stay at home. This strong bias persists in *most* [my emphasis] agricultural and industrial societies and, on that ground alone, appears to have a genetic origin. . . . My own guess is that the genetic bias is intense enough to cause a substantial division of labor even in the most free and most egalitarian of future societies. . . . Even with identical education and equal access to all professions, men are likely to continue to play a disproportionate role in political life, business and science.

We are both similar to and different from other animals. In different cultural contexts, emphasis upon one side or the other of this fundamental truth plays a useful social role. In Darwin's day, an assertion of our similarity broke through centuries of harmful superstition. Now we may need to emphasize our difference as flexible animals with a vast range of potential behavior. Our biological nature does not stand in the way of social reform. We are, as Simone de Beauvoir said, "l'être dont l'être est de n'être pas"—the being whose essence lies in having no essence.

33 So Cleverly Kind an Animal

IN *Civilization and Its Discontents,* Sigmund Freud examined the agonizing dilemma of human social life. We are by nature selfish and aggressive, yet any successful civilization demands that we suppress our biological inclinations and act altruistically for common good and harmony. Freud argued further that as civilizations become increasingly complex and "modern," we must renounce more and more of our innate selves. This we do imperfectly, with guilt, pain, and hardship; the price of civilization is individual suffering.

> It is impossible to overlook the extent to which civilization is built up upon a renunciation of instinct, how much it presupposes precisely the nonsatisfaction . . . of powerful instincts. This "cultural frustration" dominates the large field of social relationships between human beings.

Freud's argument is a particularly forceful variation on a ubiquitous theme in speculations about "human nature." What we criticize in ourselves, we attribute to our animal past. These are the shackles of our apish ancestry—brutality, aggression, selfishness; in short, general nastiness. What we prize and strive for (with pitifully limited success), we consider as a unique overlay, conceived by our rationality and imposed upon an unwilling body. Our hopes for a better future lie in reason and kindness—the mental transcendence of our biological limitations. "Build thee more stately mansions, O my soul."

Little more than ancient prejudice supports this common belief. It certainly gains no justification from science—so profound is our ignorance about the biology of human behavior. It arises from such sources as the theology of the human soul and the "dualism" of philosophers who sought separate realms for mind and body. It has roots in an attitude that I attack in several of these essays: our desire to view the history of life as progressive and to place ourselves on top of the heap (with all the prerogatives of domination). We seek a criterion for our uniqueness, settle (naturally) upon our minds, and define the noble results of human consciousness as something intrinsically apart from biology. But why? Why should our nastiness be the baggage of an apish past and our kindness uniquely human? Why should we not seek continuity with other animals for our "noble" traits as well?

One nagging scientific argument does seem to support this ancient prejudice. The essential ingredient of human kindness is altruism—sacrifice of our personal comfort, even our lives in extreme cases, for the benefit of others. Yet, if we accept the Darwinian mechanism of evolution, how can altruism be part of biology? Natural selection dictates that organisms act in their own self-interest. They know nothing of such abstract concepts as "the good of the species." They "struggle" continuously to increase the representation of their genes at the expense of their fellows. And that, for all its baldness, is all there is to it; we have discovered no higher principle in nature. Individual advantage, Darwin argues, is the only criterion of success in nature. The harmony of life goes no deeper. The balance of nature arises from interaction between competing teams, each trying to win the prize for itself alone, not from the cooperative sharing of limited resources.

How, then, could anything but selfishness ever evolve as a biological trait of behavior? If altruism is the cement of stable societies, then human society must be fundamentally outside nature. There is one way around this dilemma. Can an apparently altruistic act be "selfish" in this Darwinian sense? Can an individual's sacrifice ever lead to the perpetuation of his own genes? The answer to this seemingly contradictory proposition is "yes." We owe the resolution of this paradox to the theory of "kin selection" developed in the early 1960s by W. D. Hamilton, a British theoretical biologist. It has been stressed as the cornerstone for a biological theory of society in E. O. Wilson's *Sociobiology.* (I criticized the deterministic aspects of Wilson's speculations on human behavior in the last essay. I also praised his general theory of altruism, and continue this theme now.)

The legacy of brilliant men includes undeveloped foresight. English biologist J. B. S. Haldane probably anticipated every good idea that evolutionary theorists will invent during this century. Haldane, arguing about altruism one evening in a pub, reportedly made some quick calculations on the back of an envelope, and announced: "I will lay down my life for two brothers or eight cousins." What did Haldane mean by such a cryptic comment? Human chromosomes come in pairs: We receive one set from our mother's egg; the other from our father's sperm. Thus, we possess a paternal and a maternal copy of each gene (this is not true among males for genes located on sex chromosomes, since the maternal X chromosome is so much longer—i.e. has so many more genes—than the paternal Y chromosome; most genes on the X chromosome have no corresponding copy on the short Y). Take any human gene. What is the probability that a brother will share the same gene? Suppose that it is on a maternal chromosome (the argument works the same way for paternal chromosomes). Each egg cell contains one chromosome of each pair—that is, one half the mother's genes. The egg cell that made your brother either had the same chromosome you received or the other member of the pair. The chance that you share your brother's gene is an even fifty-fifty. Your brother shares half your genes and is, in the Darwinian calculus, the same as half of you.

Suppose, then, that you are walking down the road with three brothers. A monster approaches with clearly murderous intent. Your brothers do not see it. You have only two alternatives: Approach it and give a rousing Bronx cheer, thereby warning your brothers, who hide and escape, and insuring your own demise; or hide and watch the monster feast on your three brothers. What, as an accomplished player of the Darwinian game, should you do? The answer must be, step right up and cheer—for you have only yourself to lose, while your three brothers represent one and a half of you. Better that they should live to propagate 150 percent of your genes. Your apparently altruistic act is genetically "selfish," for it maximizes the contribution of your genes to the next generation.

According to the theory of kin selection, animals evolve behaviors that endanger or sacrifice themselves only if such altruistic acts increase their own genetic potential by benefiting kin. Altruism and the society of kin must go hand in hand; the benefits of kin selection may even propel the evolution of social interaction. While my absurd example of four brothers and a monster is simplistic, the situation becomes much more complex with twelfth cousins, four times removed. Hamilton's theory does not only belabor the obvious.

Hamilton's theory has had stunning success in explaining some persistent biological puzzles in the evolution of social behavior in the Hymenoptera—ants, bees, and wasps. Why has true sociality evolved independently at least eleven times in the Hymenoptera and only once among other insects (the termites)? Why are sterile worker castes always female in the Hymenoptera, but both male and female in termites? The answers seem to lie in the workings of kin selection within the unusual genetic system of the Hymenoptera.

Most sexually reproducing animals are diploid; their cells contain two sets of chromosomes—one derived from their mother, the other from their father. Termites, like most insects, are diploid. The social Hymenoptera, on the other hand, are haplodiploid. Females develop from fertilized eggs as normal diploid individuals with maternal and paternal sets of chromosomes. But males develop from unfertilized eggs and possess only the maternal set of chromosomes; they are, in technical parlance, haploid (half the normal number of chromosomes).

In diploid organisms, genetic relationships of sibs and parents are symmetrical: parents share half their genes with their children, and each sib (on average) shares half its genes with any other sib, male or female. But in haplodiploid species, genetic relationships are asymmetrical, permitting kin selection to work in an unusual and potent way. Consider the relationship of a queen ant to her sons and daughters, and the relationship of these daughters to their sisters and brothers:

1. The queen is related by 1/2 to both her sons and daughters; each of her offspring carries 1/2 her chromosomes and, therefore, 1/2 her genes.

2. Sisters are related to their brothers, not by 1/2 as in diploid organisms, but only by 1/4. Take any of a sister's genes. Chances are 1/2 that it is a paternal gene. If so, she cannot share it with her brother (who has no paternal genes). If it is a maternal gene, then chances are 1/2 that her brother has it as well. Her total relationship with her brother is the average of zero (for paternal genes) and 1/2 (for maternal genes), or 1/4.

3. Sisters are related to their sisters by 3/4. Again, take any gene. If it is paternal, then her sister must share it (since fathers have only one set of chromosomes to pass to all daughters). If it is maternal, then her sister has a fifty-fifty chance of sharing it, as before. Sisters are related by the average of 1 (for paternal genes) and 1/2 (for maternal genes), or 3/4.

These asymmetries seem to provide a simple and elegant explanation for that most altruistic of animal behaviors—the "willingness" of sterile female workers to forego their own reproduction in order to help their mothers raise more sisters. As long as a worker can invest preferentially in her sisters, she will perpetuate more of her genes by helping her mother raise fertile sisters (3/4 relationship) than by raising fertile daughters herself (1/2 relationship). But a male has no inclination toward sterility and labor. He would much rather raise daughters, who share all his genes, than help sisters, who share only 1/2 of them. (I do not mean to attribute conscious will to creatures with such rudimentary brains. I use such phrases as "he would rather" only as a convenient shortcut for "in the course of evolution, males who did not behave this way have been placed at a selective disadvantage and gradually eliminated.")

My colleagues R. L. Trivers and H. Hare have recently reported the following important discovery in *Science* (January 23, 1976): they argue that queens and workers should prefer different sex ratios for fertile offspring. The queen favors a 1:1 ratio of males to females since she is equally related (by 1/2) to her sons and daughters. But the workers raise the offspring and can impose their preferences upon the queen by selective nurturing of her eggs. Workers would rather raise fertile sisters (relationship 3/4) than brothers (relationship 1/4). But they must raise some brothers, lest their sisters fail to find mates. So they compromise by favoring sisters to the extent of their stronger relationship to them. Since they are three times more related to sisters than brothers, they should invest three times more energy in raising sisters. Workers invest energy by feeding; the extent of feeding is reflected in the adult weight of fertile offspring. Trivers and Hare therefore measured the ratio of female/-male weight for all fertile offspring taken together in nests of 21 different ant species. The average weight ratio—or investment ratio—is remarkably close to 3:1. This is impressive enough, but the clincher in the argument comes from studies of slave-making ants. Here, the workers are captured members of other species. They have no genetic relationship to the daughters of their imposed queen and should not favor them over the queen's sons. Sure enough, in these situations, the female/male weight ratio is 1:1—even though it is again 3:1 when workers of the enslaved species are not captured but work, instead, for their own queen.

Kin selection, operating on the peculiar genetics of haplodiploidy, seems to explain the key features of social behavior in ants, bees, and wasps. But what can it do for us? How can it help us understand the contradictory amalgam of impulses toward selfishness and altruism that form our own personalities. I am willing to admit—and this is only my intuition, since we have no facts to constrain us—that it probably resolves Freud's dilemma of the first paragraph. Our selfish and aggressive urges may have evolved by the Darwinian route of individual advantage, but our altruistic tendencies need not represent a unique overlay imposed by the demands of civilization. These tendencies may have arisen by the same Darwinian route via kin selection. Basic human kindness may be as "animal" as human nastiness.

But here I stop—short of any deterministic speculation that attributes *specific* behaviors to the possession of specific altruist or opportunist genes. Our genetic makeup permits a wide range of behaviors—from Ebenezer Scrooge before to Ebenezer Scrooge after. I do not believe that the miser hoards through opportunist genes or that the philanthropist gives because nature endowed him with more than the normal complement of altruist genes. Upbringing, culture, class, status, and all the intangibles that we call "free will," determine how we restrict our behaviors from the wide spectrum —extreme altruism to extreme selfishness—that our genes permit.

As an example of deterministic speculations based on altruism and kin selection, E.O. Wilson has proposed a genetic explanation of homosexuality (*New York Times Magazine,* October 12, 1975). Since exclusive homosexuals do not bear children, how could a homosexuality gene ever be selected in a Darwinian world? Suppose that our ancestors organized socially as small, competing groups of very close kin. Some groups contained only heterosexual members. Others included homosexuals who functioned as "helpers" in hunting or child rearing: they bore no children but they helped kin to raise their close genetic relatives. If groups with homosexual helpers prevailed in competition over exclusively heterosexual groups, then homosexuality genes would have been maintained by kin selection. There is nothing illogical in this proposal, but it has no facts going for it either. We have identified no homosexuality gene, and we know nothing relevant to this hypothesis about the social organization of our ancestors.

Wilson's intent is admirable; he attempts to affirm the intrinsic dignity of a common and much maligned sexual behavior by arguing that it is natural for some people—and adaptive to boot (at least under an ancestral form of social organization). But the strategy is a dangerous one, for it backfires if the genetic speculation is wrong. If you defend a behavior by arguing that people are programmed directly for it, then how do you continue to defend it if your speculation is wrong, for the behavior then becomes unnatural and worthy of condemnation. Better to stick resolutely to a philosophical position on human liberty: what free adults do with each other in their own private lives is their business alone. It need not be vindicated—and must not be condemned—by genetic speculation.

Although I worry long and hard about the deterministic uses of kin selection, I applaud the insight it offers for my favored theme of biological potentiality. For it extends the realm of genetic potential even further by including the capacity for kindness, once viewed as intrinsically unique to human culture. Sigmund Freud argued that the history of our greatest scientific insights has reflected, ironically, a continuous retreat of our species from center stage in the cosmos. Before Copernicus and Newton, we thought we lived at the hub of the universe. Before Darwin, we thought that a benevolent God had created us. Before Freud, we imagined ourselves as rational creatures (surely one of the least modest statements in intellectual history). If kin selection marks another stage in this retreat, it will serve us well by nudging our thinking away from domination and toward a perception of respect and unity with other animals.

Epilogue

WHERE IS DARWINISM going? What are the prospects for its second century? I claim no clairvoyance, only some knowledge of the past. But I do believe that an assessment of future direction must be tied to an understanding of what has been—particularly to the three central ingredients of Darwin's own world view: his focus on individuals as primary evolutionary agents, his identification of natural selection as the mechanism of adaptation, and his belief in the gradual nature of evolutionary change.

Did Darwin hold that natural selection acts as an exclusive agent of evolutionary change? Did he believe that all products of evolution are adaptive? During the late nineteenth century, a debate arose in biological circles over who rightly wore the title of "Darwinian." August Weismann, a rigid selectionist who granted almost no role to any other mechanism, claimed the mantle as Darwin's true descendant. G. J. Romanes, who gave Lamarck and a host of other aspirants equal billing with natural selection, demanded the title for himself. Both and neither were right. Darwin's view was pluralistic and accommodating—the only reasonable stance before such a complex world. He certainly granted overwhelming importance to natural selection (Weismann), but he did not reject an influence for other factors (Romanes).

The Weismann-Romanes debate is playing itself out again, as the two most widely discussed movements of recent years line up behind the old advocates. I suspect that Darwin's middle position will prevail again, as extreme formulations on either side retreat before the multifariousness of nature. On the one side, human "sociobiologists" are presenting a series of elaborate speculations rooted in the premise that all major patterns of behavior must be adaptive as the products of natural selection. I have heard adaptive (and even genetic) arguments for such phenomena as the inheritance of wealth and property through male lines and the higher incidence of fellatio and cunnilingus among the upper classes.

With supreme confidence in universal adaptation, sociobiologists are advocating the ultimate atomism—reduction to a level even below the apparently indivisible individual of Darwin's formulation. Samuel Butler, in a famous remark, once stated that a chicken is merely the egg's way of making another egg. Some sociobiologists take this epigram literally and argue that individuals are no more than instruments that genes use to make more genes like themselves. Individuals become temporary receptacles for the "real" units of evolution. In Darwin's world, individuals struggle to perpetuate their kind. Here genes themselves are generals in the battle for survival. In such intense combat, only the fittest win; all change must be adaptive.

Wolfgang Wickler remarks: "It follows from evolutionary theory that the genes run the individual in their own interest." I confess that I cannot regard such a statement as much more than metaphorical nonsense. I am not bothered by the false attribution of conscious purpose; this is a literary convention, and I am guilty of it myself. I am disturbed by the erroneous idea that genes are discrete and divisible particles, using the traits that they build in organisms as weapons for their personal propagation. An individual is not decomposable into independent bits of genetic coding. The bits have no meaning outside the milieu of their body, and they do not directly code for any bounded piece of morphology or any specific behavior. Morphology and behavior are not rigidly built by battling genes; they need not be adaptive in all cases.

While the sociobiologists try to out-Weismann Weismann, many molecular evolutionists take the opposite view that much evolutionary change is not only uninfluenced by selection, but truly random in direction. (In Darwin's formulation, the raw material of variation may be random, but evolutionary *change* is deterministic and directed by natural selection). The genetic code, for example, is redundant. More than one sequence of DNA yields the same amino acid. It is hard to imagine how a genetic change from one redundant sequence to another can be controlled by natural selection (since selection will "see" the same amino acid in both cases).

We may choose to regard such "invisible" genetic change as irrelevant, for if variation is not expressed in an organism's morphology or physiology, natural selection cannot act upon it. Still, if most evolutionary change were neutral in this sense (I don't believe that it is), then we would need a new metaphor for Darwinian influence. We might have to view natural selection as an epiphenomenon, touching only the few genetic variations that translate into adaptively meaningful parts of organisms—a mere surface skin on a vast sea of hidden variability.

But the challenge of molecular evolutionists is more serious than this—for they have detected more variability in proteins (i.e. in visible genetic products) than models based on natural selection should permit a population to maintain. In addition, they have inferred a strikingly regular, almost clocklike, rate for evolutionary changes in proteins over long periods of time. How can evolution work like a clock if it is directed by a deterministic process like natural selection—for intensity of selection maps rates of environmental change, and climate does not tick like a metronome. Perhaps these genetic changes are truly neutral, accumulating at random and at constant rate. The issue is not settled; copious variability and clocklike rates can arise by natural selection with the aid of some ad hoc hypotheses that may not turn out to be absurd. I only wish to argue that we have no final answers.

I predict the triumph of Darwinian pluralism. Natural selection will turn out to be far more important than some molecular evolutionists imagine, but it will not be omnipotent, as some sociobiologists seem to maintain. In fact, I suspect that Darwinian natural selection based on genetic variation has rather little to do with the very behaviors now so ardently cited in its support.

I hope that the pluralistic spirit of Darwin's own work will permeate more areas of evolutionary thought, where rigid dogmas still reign as a consequence of unquestioned preference, old habits, or social prejudice. My own favorite target is the belief in slow and steady evolutionary change preached by most paleontologists (and encouraged, admittedly, by Darwin's own preferences). The fossil record does not support it; mass extinction and abrupt origination reign. We cannot demonstrate evolution by recording gradual change in some brachiopod as we climb a hillslope. To sidestep this unpleasant truth, paleontologists have relied on the extreme inadequacy of the fossil record—all the intermediate stages are missing in a record that preserves only a few words of the few lines of the few pages left in our geological book. They have purchased their gradualistic orthodoxy at the exorbitant price of admitting that the fossil record almost never displays the very phenomenon they wish to study. But I believe that gradualism is not exclusively valid (in fact, I regard it as rather rare). Natural selection contains no statement about rates. It can encompass rapid (geologically instantaneous) change by speciation in small populations as well as the conventional and immeasurably slow transformation of entire lineages.

Aristotle argued that most great controversies are resolved at the *aurea mediocritas*—the golden mean. Nature is so wondrously complex and varied that almost anything possible does happen. Captain Corcoran's "hardly ever" is the strongest statement that a natural historian can make. A person who wants clean, definitive, global answers to the problems of life must search elsewhere, not in nature. In fact, I rather doubt that an honest search will reveal such answers anywhere. We can resolve small questions definitely (I know why the world will never see an ant 25 feet long). We do reasonably well with middle-sized questions (I doubt that Lamarckism will ever enjoy a resurgence as a viable theory of evolution). Really big questions succumb to the richness of nature—change can be directed or aimless, gradual or cataclysmic, selective or neutral. I will rejoice in the multifariousness of nature and leave the chimera of certainty to politicians and preachers.

Bibliography

Ardrey, R., 1961. *African genesis.* 1967 ed. Collins: Fontana Library.

———. 1967. *The territorial imperative.* 1969 ed. Collins: Fontana Library.

Berkner, L. V., and Marshall, L. 1964. The history of oxygenic concentration in the earth's atmosphere. *Discussions of the Faraday Society* 37: 122–41.

Bethell, T. 1976. Darwin's mistake. *Harpers* (February).

Bettelheim, B. 1976. *The uses of enchantment.* New York: A. Knopf.

Bolk, L. 1926. *Das Problem der Menschwerdung.* Jena: Gustav Fischer.

Burstyn, H. L. 1975. If Darwin wasn't the Beagle's naturalist, why was he on board. *British Journal for the History of Science* 8: 62–69.

Coon, C. 1962. *The origin of races.* New York: A. Knopf.

Darwin, C. 1859. *The origin of species.* London: John Murray. (Facsimile edition, E. Mayr (ed.), Harvard University Press, 1964.)

———. 1871. *The descent of man.* 2 vols., London: John Murray.

———. 1872. *The expression of the emotions in man and animals.* London: John Murray.

———. 1887. Autobiography. In F. Darwin (ed.), *The Life and Letters of Charles Darwin.* Vol. 1. London: John Murray.

Dybus, H. S. and Lloyd, M. 1974. The habits of 17-year

periodical cicadas (Homoptera: Cicadidae: Magicicada spp.). *Ecological Monographs* 44: 279–324.

Ellis, H. 1894. *Man and woman.* New York: Charles Scribner's Sons.

Engels, F. 1876. On the part played by labor in the transition from ape to man. In *Dialectics of Nature.* 1954 ed. Moscow: Foreign Languages Publishing House.

Eysenck, H. J. 1971. *The IQ argument: race, intelligence and education.* New York: Library Press.

Freud, S. 1930. *Civilization and its discontents.* Translated by J. Strachey. 1961 ed. New York: W.W. Norton.

Gardner, R. A., and Gardner, B. T. 1975. Early signs of language in child and chimpanzee. *Science* 187: 752–53 .

Geist, V. 1971. *Mountain sheep: a study in behavior and evolution.* Chicago: University of Chicago Press.

Gould, S. J. 1974. The evolutionary significance of "bizarre" structures: antler size and skull size in the "Irish Elk," *Megaloceros giganteus. Evolution* 28: 191–220.

Gould, S. J.; Raup, D. M.; Sepkoski, J. J., Jr.; Schopf, T. J. M.; and Simberloff, D. S. 1977. The shape of evolution—a comparison of real and random clades. *Paleobiology* 3, in press.

Gruber, H. E., and Barrett, P. H. 1974. *Darwin on man: a psychological study of scientific creativity.* New York: E. P. Dutton.

Gruber, J. W. 1969. Who was the Beagle's naturalist? *British Journal for the History of Science* 4: 266–82.

Hamilton, W. D. 1964. The genetical theory of social behavior. *Journal of Theoretical Biology* 7: 1–52.

Harris, M. 1974. *Cows, pigs, wars and witches: the riddles of culture.* New York: Random House.

Huxley, A. 1939. *After many a summer dies the swan.* 1955 ed. London, Penguin.

Huxley, J. 1932. *Problems of relative growth.* London: MacVeagh. (Reprinted as Dover paperback, 1972.)

Janzen, D. 1976. Why bamboos wait so long to flower. *Annual Review of Ecology and Systematics* 7: 347–91.

Jensen, A. R. 1969. How much can we boost IQ and scholastic achievement? *Harvard Educational Review* 39: 1–123 .

Jerison, H. J. 1973. *Evolution of the brain and intelligence.* New York: Academic Press.

Johnston, R. F., and Selander, R. K. 1964. House sparrows: rapid evolution of races in North America. *Science* 144: 548–50.

Kamin, L. 1974. *The science and politics of IQ.* Potomac, Md.: Lawrence Erlbaum Associates.

King, M. C., and Wilson, A. C. 1975. Evolution at two levels in humans and chimpanzees. *Science* 188: 107–16.

Koestler, A. 1967. *The ghost in the machine.* New York: Macmillan.

———. 1971. *The case of the midwife toad.* New York: Random House.

Kraemer, L. R. 1970. The mantle flaps in three species of Lampsilis (Pelecypoda: Unionidae). *Malacologia* 10: 225–82.

Krogman, W. M. 1972. *Child growth.* Ann Arbor: University of Michigan Press.

Lloyd, M., and Dybus, H. S. 1966. The periodical cicada problem. *Evolution* 20: 133–49.

Lockard, J. S.; McDonald, L. L.; Clifford, D. A.; and Martinez, R. 1976. Panhandling: sharing of resources. *Science* 191: 406–408.

Lombroso, C. 1911. *Crime: its causes and remedies.* Boston: Little, Brown and Co.

Lorenz, K. 1966. *On aggression.* 1967 ed. London, Methuen.

Lull, R. S. 1924. *Organic evolution.* New York: Macmillan.

MacArthur, R., and Wilson, E. O. 1967. *The theory of island biogeography.* Princeton: Princeton University Press.

Margulis, L. 1974. Five-kingdom classification and the origin and evolution of cells. *Evolutionary Biology.* 7: 45–78.

Martin, R. 1975. Strategies of reproduction. *Natural History* (November), pp. 48–57.

Mayr, E. 1942. *Systematics and the origin of species.* New York: Columbia University Press.

Montagu, A. 1961. Neonatal and infant immaturity in man. *Journal of the American Medical Association* 178: 56–57.

———(ed.). 1964. *The concept of race.* London: Collier Books.

Morris, D. 1967. *The naked ape.* New York: McGraw-Hill.

Oxnard, C. 1975. *Uniqueness and diversity in human evolution:*

morphometric studies of australopithecines. Chicago: University of Chicago Press.

Passingham, R. E. 1975. Changes in the size and organization of the brain in man and his ancestors. *Brain, Behavior and Evolution* 11: 73–90.

Pilbeam, D., and Gould, S. J. 1974. Size and scaling in human evolution. *Science* 186: 892–901.

Portmann, A. 1945. Die Ontogenese des Menschen als Problem der Evolutionsforschung. *Verhandlungen der schweizerischen naturforschenden Gesellschaft,* pp. 44–53.

Press, F., and Siever, R. 1974. *Earth.* San Francisco: W. H. Freeman.

Raup, D. M.; Gould, S. J.; Schopf, T. J. M.; and Simberloff, D. 1973 . Stochastic models of phylogeny and the evolution of diversity. *Journal of Geology* 81: 525–42.

Ridley, W. I 1976. Petrology of lunar rocks and implication to lunar evolution. *Annual Review of Earth and Planetary Sciences,* pp. 15–48.

Samuelson, P. 1975. Social Darwinism. *Newsweek,* July 7.

Schopf, J. W., and Oehler, D. Z. 1976. How old are the eukaryotes? *Science,* 193:47–49.

Schopf, T. J. M. 1974. Permo-Triassic extinctions: relation to sea-floor spreading. *Journal of Geology* 82: 129–43.

Simberloff, D. S. 1974. Permo-Triassic extinctions: effects of area on biotic equilibrium. *Journal of Geology* 82: 267–74.

Stanley, S. 1973. An ecological theory for the sudden origin of multicellular life in the Late Precambrian. *Proceedings of the National Academy of Sciences* 70: 1486–89.

———. 1975. Fossil data and the Precambrian-Cambrian evolutionary transition. *American Journal of Science* 276: 56–76.

Tiger, L., and Fox, R. 1971. *The imperial animal.* New York: Holt, Rinehart and Winston.

Trivers, R., and Hare, H. 1976. Haplodiploidy and the evolution of the social insects. *Science* 191: 249–63 .

Ulrich, H.; Petalas, A.; and Camenzind, R. 1972. Der Generationswechsel von *Mycophila speyeri* Barnes, einer Gallmücke mit paedogenetischer Fortpflanzung. *Revue suisse de zoologie* 79 (supplement): 75–83.

Velikovsky, I. 1950. *Worlds in collision.* 1965 ed. New York: Delta.

———. 1955. *Earth in upheaval.* 1965 ed. New York: Delta.

Wegener, A. 1966. *The origin of continents and oceans.* New York: Dover.

Welsh, J. 1969. Mussels on the move. *Natural History* (May): 56–59.

Went, F. W. 1968. The size of man. *American Scientist* 56: 400–413.

Whittaker, R. H. 1969. New concepts of kingdoms of organisms. *Science* 163: 150–60.

Wilson, E. O. 1975. *Sociobiology.* Cambridge, Mass.: Harvard University Press.

———. 1975. Human decency is animal. *New York Times Magazine,* Oct. 12.

Young, J. Z. 1971. *An introduction to the study of man.* Oxford: Oxford University Press.

Index